AUTOSAR Fundamentals and Applications

Establishing a solid foundation for automotive software design with AUTOSAR

Hossam Soffar

AUTOSAR Fundamentals and Applications

Copyright © 2024 Packt Publishing

All rights reserved. No part of this book may be reproduced, stored in a retrieval system, or transmitted in any form or by any means, without the prior written permission of the publisher, except in the case of brief quotations embedded in critical articles or reviews.

Every effort has been made in the preparation of this book to ensure the accuracy of the information presented. However, the information contained in this book is sold without warranty, either express or implied. Neither the author, nor Packt Publishing or its dealers and distributors, will be held liable for any damages caused or alleged to have been caused directly or indirectly by this book.

Packt Publishing has endeavored to provide trademark information about all of the companies and products mentioned in this book by the appropriate use of capitals. However, Packt Publishing cannot guarantee the accuracy of this information.

Group Product Manager: Preet Ahuja

Publishing Product Manager: Surbhi Suman

Book Project Manager: Ashwin Dinesh Kharwa

Senior Editor: Apramit Bhattacharya

Technical Editor: Nithik Cheruvakodan

Copy Editor: Safis Editing

Proofreader: Apramit Bhattacharya

Indexer: Rekha Nair

Production Designer: Nilesh Mohite

DevRel Marketing Coordinator: Rohan Dobhal

First published: December 2024

Production reference: 1221124

Published by Packt Publishing Ltd.

Grosvenor House

11 St Paul's Square

Birmingham

B3 1RB, UK

ISBN 978-1-80512-087-2

www.packtpub.com

This book is humbly dedicated to all the extraordinary individuals who have supported and inspired me throughout my journey. To my beloved family: your unwavering love and encouragement have been the foundation of all my endeavors. Thank you for your endless support and belief in me.

To my colleagues, past and present, from Valeo Egypt, Elektrobit, FEV, and now Plus AI: your creativity, dedication, and collaboration have been invaluable. Working alongside such talented and innovative minds has been a true privilege, and I am grateful for every moment we have shared. You have taught me countless lessons and enriched my career in ways I could never have imagined.

I also dedicate this book to all the unfortunate souls in the world. May you find peace and justice one day. Your resilience and strength inspire us to strive for a better, more equitable world.

Contributors

About the author

Hossam Soffar is a seasoned automotive embedded software engineer with over 12 years of experience in the automotive software industry. He has acquired substantial expertise in AUTOSAR through his roles in development and consulting for automotive ECUs, collaborating with an extensive network of OEMs and Tier 1 and Tier 2 suppliers. He has been involved in debugging embedded software, establishing software architectures, improving software performance, and managing software requirements and architectural design, with a particular focus on safety, security, and real-time performance. Committed to lifelong learning, Hossam values the opportunity to share his knowledge and insights with colleagues and peers in the field.

About the reviewers

Shriram Gobichettipalayam Ramalingam is an automotive embedded software architecture specialist with 14+ years of experience in the automotive domain. He has contributed to designing in-vehicle network architecture, classic and adaptive AUTOSAR software stack configuration, and the development of various OEMs.

I would like to thank my manager, Mr. Sandeep, who motivated me to expand my experience beyond work. Thank you for breaking stereotypes and being understanding and supportive in all activities.

Dan Stroud is a senior solutions architect for AWS, specializing in the automotive sector. He gained his CEng in 2016 and has an MSc in systems engineering management (UCL) and a BSc in software engineering (University of Portsmouth). His 17+ years of experience in software development and architecture includes flight controls architecture (777x), leading the design of a new EV platform architecture for an automotive OEM, heavily focused around the AUTOSAR framework, and now specializing in helping automotive customers to deliver connected features and accelerate innovation through the cloud.

The automotive industry has seen a fundamental shift, with innovation now being driven by software. With resources limited and expectations high, it's technologies such as AUTOSAR that allow developers to focus on features, rather than re-inventing the platform wheel. I feel privileged to have experienced this as an OEM and I'm excited to see how connectivity and the power of the cloud drive the next cycle of innovation.

Alaa Mahran is a senior embedded software engineer with over 10 years of experience in automotive software, specializing in AUTOSAR. He currently works at Webasto SE, where he leads software architecture and bootloader development, focusing on safety-critical and cybersecurity-compliant projects. Alaa has also worked with Valeo on ADAS applications, contributing to enhancing automotive safety and driver assistance features.

I want to express my appreciation for my father, who taught me to keep an eye out for details and strive for excellence and continuous improvement.

Table of Contents

Preface — xiii

Free Benefits with Your Book — xviii

Part 1: Introduction – The Genesis and Framework of AUTOSAR

1

Exploring the Genesis and Objectives of AUTOSAR — 3

Evolution of the automotive industry	3	Case study – Developing an ECU in the AUTOSAR framework	13
What is an ECU?	5		
Introducing automotive software development	6	Understanding the AUTOSAR standards	13
Understanding traditional software development	7	Software architecture and design	17
Case study – Replacing an MCU and exchanging microcontroller drivers	9	Layers	17
Introducing the AUTOSAR framework	9	Stacks	19
Impact on traditional software development	11	Summary	20
How were these goals achieved?	12	Questions	21

2

Introducing the AUTOSAR Software Layers — 23

Understanding layered software architecture in AUTOSAR	24	Understanding the function of the RTE	28
Application layer	25	Basic software layer	31
Runtime environment	27	Service layer	31
Understanding the VFB concept	27	ECU abstraction layer	33
		Microcontroller abstraction layer	34

Complex drivers	36	BSW – Service layer	40
Interfaces	**37**	BSW – ECU abstraction layer	41
Case study – Developing a BMS ECU	**39**	BSW – MCAL	41
Application layer	40	**Summary**	**42**
RTE	40	**Exercise**	**42**

3

AUTOSAR Methodology and Data Exchange Formats 43

Introducing the AUTOSAR methodology	**43**	Code Generation	49
Understanding the AUTOSAR methodology	**45**	**Using templates for data exchange**	**51**
		Conformance classes for AUTOSAR	**54**
System Configuration	45	**Summary**	**54**
ECU Configuration	48	**Exercise**	**55**

Part 2: Investigating the Building Blocks of AUTOSAR

4

Working with Software Components and RTE 59

Understanding SWCs	**60**	Introducing compositions	73
Example – Throttle controller	61	Connector types	74
Types of AUTOSAR SWCs	62	**Understanding the significance of the RTE**	**76**
Elements of SWC	63		
Exploring AUTOSAR datatypes	**67**	Generation of RTE	77
Modeling an SWC	**69**	**Summary**	**81**
What is an AUTOSAR authoring tool?	69	**Questions**	**82**
Communication between the SWCs	**72**		

5

Designing and Implementing Events and Interfaces — 83

Introducing the communication model	83	How do ports, interfaces, and events interact with each other?	103
Explaining the information flow	84	Events for temperature monitoring example	108
Sender-receiver communication	87	Summary	109
Client-server communication	95	Questions	109
Rethinking temperature sensor design	100		
Understanding events in AUTOSAR	102		

6

Getting Started with the AUTOSAR Operating System — 111

A brief overview of the AUTOSAR OS	111	Interrupts	119
Introduction to real-time operating systems	112	Events	120
Introduction to OSEK	113	Scheduling	122
Configuring the AUTOSAR OS	114	Task trigger mechanisms	124
		OS resources	128
Exploring the AUTOSAR OS architecture	115	Hooks	129
Tasks	116	Summary	130
		Questions	131

7

Exploring the Communication Stack — 133

What is the COM stack?	134	Network management	157
COM stack overview	134	Key modules in network management	157
An overview of basic COM concepts	137	Summary	160
COM module	141	Questions	160
PDU-Router (PduR) Module	149		
CAN modules	151		
Exploring an example of sending a CAN message	154		

Part 3: Beyond Fundamentals – Advanced AUTOSAR Concepts

8

Securing the AUTOSAR System with Crypto and Security Stack — 163

Introduction to automotive security	163
Fundamentals of automotive security	164
Security in AUTOSAR	165
Differentiating between safety and security	166
Security stack architecture in AUTOSAR	**167**
Cryptographic techniques	167
Crypto service manager module	169
Crypto Interface module	171
Crypto module	171
Other helper modules to achieve security	173
Crypto Stack use cases and examples	**174**
Data encryption	174
Secure communication	177
Secure diagnostics	179
Secure boot in AUTOSAR	180
Summary	**182**
Questions	**182**

9

Dealing with Memory and Mode Management — 183

Memory stack in AUTOSAR	**184**
Significance of non-volatile memory in AUTOSAR	184
Understanding the memory stack	184
Storage objects	186
Use case	190
Mode management	**192**
Role and functionality	192
Use case for the BSWM – Vehicle shutdown process	196
Summary	**198**
Questions	**198**

10

Wrapping Up and Extending Knowledge with a Use Case — 199

Overview of diagnostics	**200**
Diagnostic Communication Manager (DCM) in detail	201
Diagnostic Event Manager	202
Default Error Tracer (DET)	204
Failure Management	205
Overview of time synchronization	**206**

Synchronized Time-Base Manager (StbM) module	207	What's next?	213
CAN Time Synchronization (CanTSyn)	209	Case study – Development of an autonomous parking assist system	214
How to read the standards	**210**	Application layer (SWCs)	217
Key AUTOSAR documents	211	AUTOSAR BSW modules	218
		Summary	**219**

Index 221

Other Books You May Enjoy 232

Preface

Embedded software teams use the AUTOSAR framework to standardize interfaces between application software and basic vehicular functions. This standardization helps develop and deploy software rapidly and efficiently for automotive **electronic control units** (**ECUs**) and ensures more effective project management.

Many engineers find AUTOSAR challenging due to its complexity owing to the difficulty of grasping its concepts at the beginning. The intricate architecture and extensive standards can lead to confusion, making it hard for beginners to see the full picture and understand how everything fits together. Additionally, beginner-friendly resources are limited. This book addresses these challenges with clear explanations, key concepts, and practical advice to help **beginners and intermediate-level engineers** confidently navigate the AUTOSAR landscape.

Automotive ECUs are safety-critical and hard real-time systems, and this book will provide you with the foundational knowledge and skills needed to participate in their development and manage AUTOSAR projects more effectively.

With a focus on the practical application of AUTOSAR, this book dives deep into the AUTOSAR framework and architecture and shows how to implement it in the development of automotive electronic systems using best practices and real-world use cases.

The book begins with an overview of the goals and objectives of AUTOSAR and then discusses its layered architecture, including the different AUTOSAR stacks, components and modules, and internal and external communication mechanisms. You'll discover how to configure, design, and integrate AUTOSAR software components in a **run-time environment** (**RTE**) and understand the principles of diagnostics, security, and **real-time operating system** (**RTOS**) for developing high-quality, secure, and efficient ECUs.

After reading this book, you'll have detailed insights into AUTOSAR and be skilled in building, implementing, and managing complex automotive systems confidently and efficiently.

Who this book is for

This book is designed for embedded software engineers and any software developer or software architect who works with or plans to work with automotive systems but has minimal or no knowledge of AUTOSAR. It serves as a valuable reference for project managers, students, and researchers who seek to learn about AUTOSAR and its applications or understand its main concepts. A background knowledge of software development processes and C programming will be beneficial.

What this book covers

Chapter 1, *Exploring the Genesis and Objectives of AUTOSAR*, introduces the origins and goals of the AUTOSAR standard. It explains the foundational principles and motivations behind its development, offering a comprehensive understanding of its objectives.

Chapter 2, *Introducing the AUTOSAR Software Layers*, explores the essential layers of AUTOSAR architecture, including the application layer, RTE, service layer, ECU abstraction layer, and MCAL. It highlights the design and implementation of modular and compatible software components, using a BMS ECU as a case study to illustrate practical applications. This chapter provides foundational knowledge for understanding how these layers interact and support automotive software development.

Chapter 3, *AUTOSAR Methodology and Data Exchange Formats*, outlines the AUTOSAR development methodology, emphasizing the independence of software component implementation from ECU configuration. It introduces AUTOSAR templates for data exchange and covers system design, modeling, code generation, and configuration. The chapter also explains conformance classes, detailing the essential BSW modules required for compliance.

Chapter 4, *Working with Software Components and RTE*, explores the structure, functionality, and types of AUTOSAR software components, including application software components and complex device drivers. It explains software component communication via ports, the role of runnable entities, and triggering conditions. The chapter also highlights the importance of the runtime environment and its connection with the **virtual function bus** (**VFB**), providing a foundation for developing complex automotive systems.

Chapter 5, *Designing and Implementing Events and Interfaces*, unravels the complexities of events and interfaces in the AUTOSAR framework. It emphasizes their role in data transitions, real-time responses, and safety. The chapter covers advanced communication models, including sender-receiver and client-server interfaces, and synchronous versus asynchronous communication, using a car's temperature monitoring system as an example.

Chapter 6, *Getting Started with the AUTOSAR Operating System*, examines the AUTOSAR **operating system** (**OS**), its architecture, RTOS, and the OSEK standard. It highlights priority-based scheduling, fast interrupt processing, and inter-task communication. The chapter also covers task management, synchronization, and resource allocation, providing essential knowledge for developing efficient and reliable automotive software systems.

Chapter 7, *Exploring the Communication Stack*, covers the COM module's role in data exchange, the significance of signals, and the PDUR module's function in routing and transforming data. Using the CAN stack as an example, it provides insights into the key components and mechanisms of AUTOSAR communication.

Chapter 8, Securing the AUTOSAR System with Crypto and Security Stack, focuses on automotive cybersecurity within the AUTOSAR framework. It highlights the importance of security and potential risks. The chapter explores the AUTOSAR crypto stack, including its core components and their roles in cryptographic operations. It also covers **Secure Onboard Communication (SecOC)** for ensuring data confidentiality and integrity between ECUs, emphasizing the importance of secure coding practices and regulatory compliance.

Chapter 9, Dealing with Memory and Mode Management, covers the architecture and functionalities of the NVM stack, focusing on data storage and retrieval through the **Non-Volatile Memory Manager (NVMM)** module. It provides practical insights into configuring NVM, discussing storage objects, block management, and error handling. Additionally, it discusses **Basic Software Mode Management (BSWM)**, illustrating its role in managing different operating modes within automotive ECUs to ensure seamless operation and reliability.

Chapter 10, Wrapping Up and Extending Knowledge with a Use Case, concludes our exploration of AUTOSAR, emphasizing its importance in automotive software engineering. It recaps key concepts, including AUTOSAR architecture, SWS, TPS, RS, RTE, and BSW specifications, and presents a use case for designing a real-time control system within an automotive ECU. This chapter serves as a foundation, encouraging engineers to continue studying AUTOSAR specifications, gain hands-on experience, and stay updated with industry trends to master this powerful standard.

To get the most out of this book

You should have a basic understanding of automotive software development and embedded systems. Familiarity with RTOS and general software engineering principles will be beneficial. Additionally, a foundational knowledge of communication protocols and basic cybersecurity concepts will help in grasping the more advanced topics covered in the chapters.

If you are using the digital version of this book, we advise you to type the code yourself. Doing so will help you avoid any potential errors related to the copying and pasting of code.

Conventions used

There are a number of text conventions used throughout this book.

`Code in text`: Indicates code words in text, database table names, folder names, filenames, file extensions, pathnames, dummy URLs, user input, and Twitter handles. Here is an example: "The function then reads a value and stores it in the `receivedData` variable."

A block of code is set as follows:

```
#include "Rte_Receiver.h"
void ReceiverCounterFunction(void) {
    UInt32 receivedData;
    Rte_Read_ReceiverInterface_ReceiverInput(&receivedData);
     // Process received data, e.g., control an actuator based on the input
    // In this example only odd counter value is sent
if (receivedData % 2)   {
    Rte_Write_ReciverModule_CounterSig(receivedData)
    }
}
```

Bold: Indicates a new term, an important word, or words that you see onscreen. For instance, words in menus or dialog boxes appear in **bold**. Here is an example: "**AUTomotive Open System ARchitecture (AUTOSAR)** is a standard for the development of automotive electronic systems. "

> **Tips or important notes**
> Appear like this.

Get in touch

Feedback from our readers is always welcome.

General feedback: If you have questions about any aspect of this book, email us at `customercare@packtpub.com` and mention the book title in the subject of your message.

Errata: Although we have taken every care to ensure the accuracy of our content, mistakes do happen. If you have found a mistake in this book, we would be grateful if you would report this to us. Please visit `www.packtpub.com/support/errata` and fill in the form.

Piracy: If you come across any illegal copies of our works in any form on the internet, we would be grateful if you would provide us with the location address or website name. Please contact us at `copyright@packt.com` with a link to the material.

If you are interested in becoming an author: If there is a topic that you have expertise in and you are interested in either writing or contributing to a book, please visit `authors.packtpub.com`

Share Your Thoughts

Once you've read *AUTOSAR Fundamentals and Applications*, we'd love to hear your thoughts! Scan the QR code below to go straight to the Amazon review page for this book and share your feedback.

https://packt.link/r/1-805-12087-5

Your review is important to us and the tech community and will help us make sure we're delivering excellent quality content.

Free Benefits with Your Book

This book comes with free benefits to support your learning. Activate them now for instant access (see the "*How to Unlock*" section for instructions).

Here's a quick overview of what you can instantly unlock with your purchase:

PDF and ePub Copies

Next-Gen Web-Based Reader

- Access a DRM-free PDF copy of this book to read anywhere, on any device.
- Use a DRM-free ePub version with your favorite e-reader.

- **Multi-device progress sync**: Pick up where you left off, on any device.
- **Highlighting and notetaking**: Capture ideas and turn reading into lasting knowledge.
- **Bookmarking**: Save and revisit key sections whenever you need them.
- **Dark mode**: Reduce eye strain by switching to dark or sepia themes

How to Unlock

Scan the QR code (or go to packtpub.com/unlock). Search for this book by name, confirm the edition, and then follow the steps on the page.

Note: Keep your invoice handy. Purchases made directly from Packt don't require one

Part 1: Introduction – The Genesis and Framework of AUTOSAR

This part provides a foundational understanding of AUTOSAR, its origins, and its layered architecture. It introduces the key principles, motivations, and methodologies that underpin the AUTOSAR framework, offering insights into how data exchange formats facilitate seamless communication within the system. You will gain a solid grasp of the fundamental concepts necessary for developing AUTOSAR-compliant automotive software.

This part has the following chapters:

- *Chapter 1, Exploring the Genesis and Objectives of AUTOSAR*
- *Chapter 2, Introducing the AUTOSAR Software Layers*
- *Chapter 3, AUTOSAR Methodology and Data Exchange Formats*

1

Exploring the Genesis and Objectives of AUTOSAR

AUTomotive Open System ARchitecture (**AUTOSAR**) is a standard for the development of automotive electronic systems. It provides a common software architecture for **electronic control units** (**ECUs**) in vehicles, allowing for the easier integration and development of new features. It is a partnership of major automotive manufacturers and suppliers, and its goal is to improve the overall efficiency and flexibility of the automotive software development process.

In this first chapter, we will discuss the motivation behind the development of AUTOSAR, the organization of the partnership, and its aims and objectives. In this chapter, we will cover the following main topics:

- Evolution of the automotive industry
- Introducing the AUTOSAR framework
- Understanding the AUTOSAR standards
- Software architecture and design

> **Free Benefits with Your Book**
> Your purchase includes a free PDF copy of this book along with other exclusive benefits. Check the *Free Benefits with Your Book* section in the Preface to unlock them instantly and maximize your learning experience.

Evolution of the automotive industry

Over the past few decades, the automobile industry has evolved from a simple means of transportation to a complex machine that resembles a smartphone on wheels. This transformation is due to the integration of advanced technologies and the adoption of a more sophisticated approach to design and development.

The evolution of the car can be traced back to the early 20th century when the first automobiles were developed. These vehicles were simple and utilitarian, designed primarily for transportation from point A to point B. However, as technology advanced, so did the car. In the 1950s and 1960s, we saw the emergence of advanced safety features such as seat belts, airbags, and anti-lock brakes. By the 1980s, we began to see the introduction of onboard computers, which enabled more advanced engine management and diagnostics. In the 1990s, vehicles saw the integration of more sophisticated electronic systems, such as **electronic stability control** (**ESC**) and **advanced driver assistance systems** (**ADAS**).

In the 21st century, the car has undergone a massive transformation. Today's vehicles are equipped with advanced features that were once reserved for high-end luxury cars.

The percentage of car production costs attributed to electronic control systems and automotive software has been consistently rising over the years. This upward trend is clearly depicted in the Statista data (https://www.statista.com/statistics/277931/automotive-electronics-cost-as-a-share-of-total-car-cost-worldwide/) shown in the following figure, which has been monitoring this development since 1970:

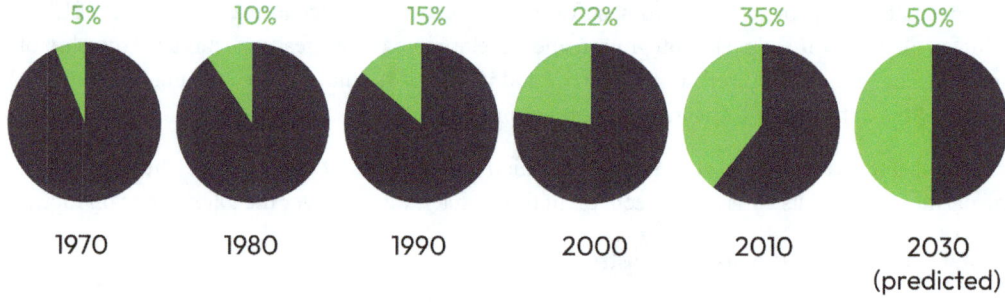

Figure 1.1 – Electronics system as percent of total car cost

This increasing complexity of automotive systems has presented a challenge for the industry in terms of software development.

The evolution of the car into a smart, connected, and autonomous machine is driven by several factors, which include the following:

- The growing demand for advanced safety features
- The need for more efficient and environmentally friendly transportation
- The desire for a more convenient and connected driving experience

The adoption of advanced technologies such as sensors, software systems, and connectivity has enabled car manufacturers to deliver on these demands, creating a new era of smart, autonomous, and connected vehicles.

With the integration of new technologies such as ADAS and connected car features, the amount of software that needs to be developed and integrated into vehicles has grown significantly.

The comparison to the Apollo mission highlights the significant increase in complexity of modern cars. While the Apollo spacecraft had only a limited number of systems that needed to be managed, modern cars can contain up to 100 or more ECUs, sensors, and actuators, all of which need to communicate seamlessly within very tight time constraints with one another to ensure proper functioning. Additionally, modern cars are highly connected devices that require sophisticated software and networking capabilities, further adding to their complexity. This increased complexity allows modern cars to offer advanced features and functionality but also requires more sophisticated maintenance and repair processes.

Having discussed the evolution and complexity of automotive software, let's shift our focus to one of the essential components that enable modern cars to function effectively – the ECU.

What is an ECU?

Before we move any further, we need to understand what an automotive ECU is. This is a computer – comprising a **printed circuit board** (**PCB**) with a microcontroller and various electronic components – that controls various functions in a vehicle. These functions may include engine management, transmission control, climate control, power steering, and brakes. Here are some examples of automotive ECUs:

- **Engine control module** (**ECM**): The ECM is responsible for managing the engine's performance, including fuel injection, ignition timing, and emissions control.
- **Transmission control module** (**TCM**): The TCM manages the operation of the transmission, including gear selection, shift timing, and torque converter lock-up.
- **Body control module** (**BCM**): The BCM controls various functions related to the vehicle's body and interior, such as lighting, climate control, door locks, and audio systems.
- **Anti-lock braking system** (**ABS**) **control module**: The ABS control module manages the operation of the ABS, which helps to prevent skidding and maintain control of the vehicle during braking.
- **Battery management system** (**BMS**): The BMS's primary function is to monitor, control, and optimize the performance of the vehicle's battery pack. It also ensures all battery cells within the pack are charged and discharged uniformly, preventing the overcharging of certain cells and maximizing the overall battery capacity.

Some examples of these components are shown in the following figure:

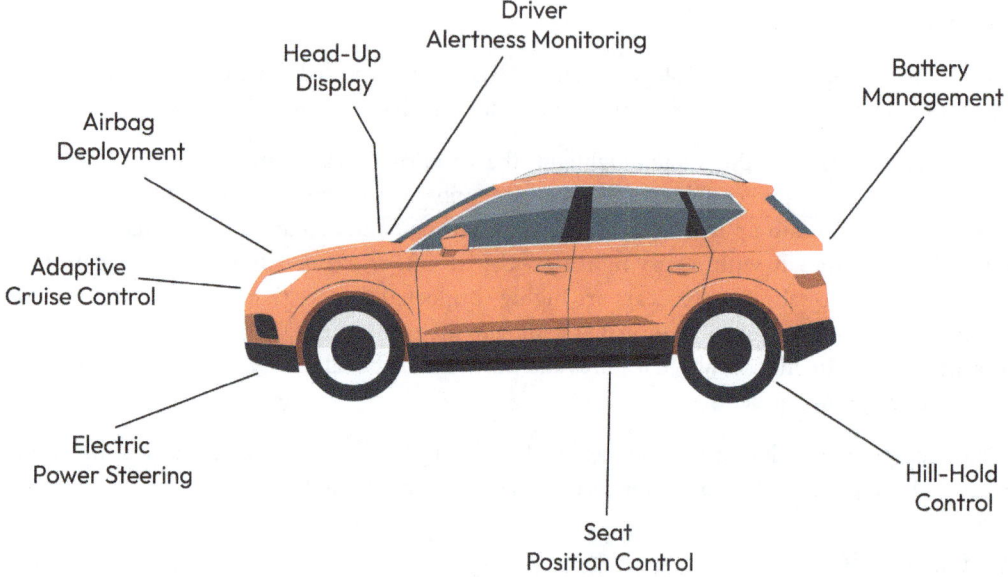

Figure 1.2 – Examples of ECUs in a vehicle

Overall, automotive ECUs play a critical role in the operation of modern vehicles, providing precise control over various systems and ensuring optimal performance, efficiency, and safety. As automotive ECUs rely heavily on complex software to perform their functions, we first need to understand the software development aspect to comprehend the nuances of ECU operation and design.

Introducing automotive software development

Automotive software development is a critical component of the continued innovation and success of the automotive industry. It involves creating and maintaining software systems used in various types of automobiles, cars, trucks, buses, and other automobiles. These software systems are responsible for multiple tasks, such as engine management, navigation, entertainment, and safety features. Therefore, engineers in this field must have expertise in embedded systems, real-time programming, control systems, and communication protocols to create reliable and safe software systems.

It is a highly specialized field that requires close collaboration with other members of the automotive development team, such as electrical and mechanical engineers and quality assurance specialists, to ensure seamless integration of the software systems into the vehicle and meet the end users' needs. Clean software architecture principles can help address the challenges of this complex field by creating a system that is easy to maintain, modify, and evolve while being resilient to change.

> **Note**
> **Clean architecture** in software design refers to a structured approach that prioritizes clarity, separation of concerns, and maintainability. It emphasizes the organization of code in a way that minimizes dependencies, allowing for easy modifications and testing. Clean architecture fosters systems that are adaptable, scalable, and easy to comprehend.

It's a challenging field but plays a critical role in the continuous success and innovation of the automotive industry. Before we discuss advancements in this field, let's first understand traditional automotive software development.

Understanding traditional software development

Traditional automotive software development involves a wide variety of ECUs with different hardware and software, which can make it difficult to ensure that all components work together efficiently. Each supplier has its own software architecture definitions, development methodology, and interfaces for ECUs, resulting in fragmented and non-standardized **software components** (**SWCs**) across the automotive industry. This approach had several limitations, including the following:

- **Limited reusability**: SWCs developed for one vehicle or system may not be reusable in another, leading to increased development costs and a longer time to market, if a similar functionality is required on another platform.
- **Integration time**: The integration of different SWCs can be a time-consuming and expensive process, particularly if the components were not designed to work together from the beginning.
- **Time to market**: The time to market for new vehicles and features can be long and costly, particularly when traditional automotive software development methods are used. This can lead to missed opportunities and lost revenue for manufacturers and suppliers.
- **Complex supply chain**: The rising complexity of software implementations is closely linked to the increasing complexity of supply chains. In this context, software developers design their components based on the requirement definitions provided by **original equipment manufacturers** (**OEMs**) or Tier 1 suppliers, who are responsible for their integration at a later stage.
- **Rigidity**: Automotive software is often monolithic and inflexible. Also, it is very hard to adapt to changing requirements and technologies.

Traditional automotive software development was fragmented, non-standardized, and costly, making it challenging to develop high-quality software and meet the growing demands of the automotive industry. An example of this type of non-standardized architecture is shown in the following figure:

8 Exploring the Genesis and Objectives of AUTOSAR

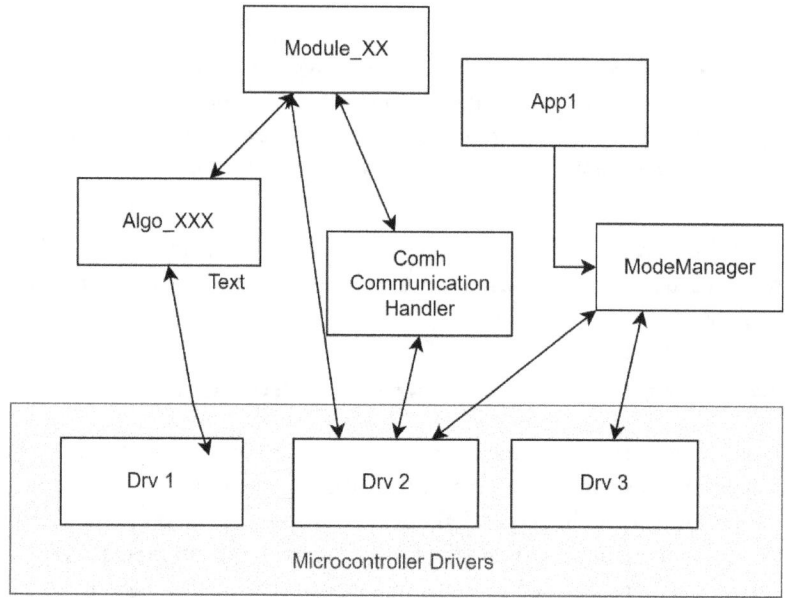

Figure 1.3 – Example of non-standardized software architecture

Thus, AUTOSAR was introduced to address these limitations and promote more efficient, effective, and standardized automotive software development. In the following section, we discuss a case study to illustrate this point.

> **Note on the evolution of automotive software standardization**
>
> There were early efforts to standardize automotive software both within individual companies and through collaborations between various entities, such as OSEK/VDX and HIS. These initiatives aimed to address specific aspects of software architecture, such as operating systems and diagnostics. Despite these efforts, they were often narrow in focus and lacked the integration needed for modern vehicle systems. This led to the development of AUTOSAR, a comprehensive standard that addresses all layers of automotive software architecture, enabling better scalability, interoperability, and reusability across different manufacturers and vehicle platforms.

Case study – Replacing an MCU and exchanging microcontroller drivers

Let's consider an example of a company wanting to upgrade an ECU, which was typically designed and implemented using proprietary software and hardware architectures, with little or no standardization across different car manufacturers. The microcontroller deployed in an ECU was typically bespoke and tailored to the car manufacturer and the specific model, and any alteration to it would necessitate extensive modifications to both the hardware and software aspects.

Suppose the company wants to replace the microcontroller of the ECU with a more advanced microcontroller. In that case, they would need to design a new hardware board that is compatible with the new microcontroller, which would likely involve changing the pin assignments and other circuitry.

With a similar architecture to that shown in *Figure 1.3*, most of the software would need to be rewritten or modified to adapt to the new microcontroller's peripheral interfaces, and architecture. This would require a significant investment in time, money, and resources, depending on the complexity of the ECU.

In summary, prior to the advent of AUTOSAR, altering the microcontroller of an automotive ECU represented a formidable task, necessitating considerable technical knowledge and regulatory expertise. The lack of standardization across automotive manufacturers compounded the issue, making it difficult to devise a universal solution. Moreover, the intricate nature of the software and hardware involved rendered any attempts to upgrade or modify an ECU a substantial challenge, requiring significant effort and resources. With the advent of AUTOSAR, there was a paradigm shift as it allowed software to be abstracted from not just the microcontroller but the entirety of the ECU and vehicle architecture. This enables developers to write applications that communicate with other software, fully abstracted from aspects such as the ECU architecture, endianness, bus architecture, signal packing and protocol, and vehicle gateways.

It's worth noting that our focus has primarily been on the benefits of AUTOSAR in relation to microcontroller replacement in this context. However, the benefits of AUTOSAR are far more comprehensive. Beyond facilitating microcontroller substitution, AUTOSAR's broad reach positively affects numerous other aspects of automotive software and hardware, making it a more efficient and flexible solution in the realm of automotive technology.

Introducing the AUTOSAR framework

AUTOSAR is an open and standardized software architecture for the automotive industry. It was developed by a consortium of automotive manufacturers, suppliers, and tool developers. The aim is to create an industry standard for automotive software architectures that is open and accessible to all. The standard is designed to meet the technical goals of the automotive software industry: **modularity**, **scalability**, **transferability**, and **function reusability**.

The main objective of AUTOSAR is to develop a **standard architecture** that can be used across different automotive domains, such as powertrain, chassis control, body, and safety. This standardization aims to enable SWCs from different suppliers to work together seamlessly, reduce development costs, and facilitate the reusability of SWCs. It also helps in managing the increasing complexity of electrical and electronic systems, as well as ensuring their quality and reliability.

Here's how the different components of the ecosystem work:

- **OEMs**: They are responsible for setting ECU software requirements and choosing Tier 1 suppliers to deliver the hardware and SWCs. By adopting AUTOSAR, OEMs ensure compliance with safety and regulatory standards while promoting modular, reusable, and scalable software development for seamless integration and compatibility across various automotive systems and suppliers.

- **Tier 1 suppliers**: A Tier 1 supplier is a company that directly supplies components or systems. It can be hardware or software to an OEM for use in vehicle production. These suppliers are considered at the top of the automotive supply chain and are responsible for providing high-quality and reliable components that meet the OEM's specifications and requirements. Examples of Tier 1 suppliers in the automotive industry include companies that provide engines, transmissions, braking systems, steering systems, and electronics components such as infotainment systems and ECUs.

- **Standard software vendors**: Standard software vendors provide AUTOSAR-compliant software modules, such as communication stacks and diagnostic services, that can be easily integrated and interchanged with other SWCs. The standard software is developed following the AUTOSAR standards and can be used by Tier 1 suppliers to build more complex software modules.

- **Semiconductor manufacturers**: Semiconductor manufacturers in the automotive industry provide the electronic components. They ensure that future hardware and software needs of the automotive industry are met.

The AUTOSAR standard is developed and maintained through a collaborative effort involving its partners, who ensure that it remains relevant and up to date by considering the necessary use cases to support the roadmaps of users. Partners are grouped based on their membership type, with varying levels of involvement in the standard's development, implementation, and usage. This approach encourages diverse stakeholder participation and reflects the needs and perspectives of the entire automotive ecosystem. The collaborative nature of AUTOSAR has been instrumental in its success, enabling it to become the industry standard for automotive software architectures. The main categories are as follows:

- **Core partners**: The core partners are BMW Group, Bosch, Continental, Daimler AG, Ford, General Motors, PSA Group, Toyota, Volkswagen Group, and Volvo Group. They were the initial members of the partnership and provided the funding and resources required to develop the AUTOSAR standard.

- **Premium partners**: A group of companies who are members of the AUTOSAR development partnership and have made significant contributions to the development and promotion of the AUTOSAR standard. Premium partners have a higher level of involvement and influence in the partnership than regular members and benefit from additional collaboration opportunities and early access to the latest AUTOSAR specifications and releases.

- **Development partners**: Development partners play an important role in the partnership by sharing their knowledge, expertise, and resources to help shape the future of the automotive industry.

- **Associate partners**: Associate partners have a lower level of involvement than core partners and premium partners, but still benefit from collaboration opportunities and access to the latest AUTOSAR specifications and release.

In summary, AUTOSAR enables all the stakeholders in the ECU development process to work together effectively by providing a common language and standardized framework. This promotes interoperability, scalability, and reuse of SWCs across different car manufacturers and reduces development time and costs while improving software quality and reliability.

Impact on traditional software development

AUTOSAR addresses the limitations of traditional automotive software development by providing a standardized approach. By promoting modularity, standardization, and scalability, AUTOSAR has made it easier to develop high-quality software that meets the increasingly demanding needs of the automotive industry. The success of AUTOSAR is evident in its widespread adoption as the industry standard for automotive software architectures.

Here are some ways in which AUTOSAR addresses the limitations of traditional software development:

- **Common platform**: It provides a common platform and architecture for software development, enabling collaboration between different suppliers and manufacturers.

- **Modular design**: It promotes modularity and standardization, making it easier to reuse SWCs across different projects and adapt to changing requirements and technologies.

- **Cost-effective integration**: The standardized approach to software development that AUTOSAR takes has made it easier and more cost-effective to integrate different SWCs.

- **Safety and security**: The framework for implementing safety and security concepts in automotive software helps to ensure the safety and security of vehicles on the road.

- **Interoperability**: It defines a common methodology for communication and enables ECUs from different suppliers to interoperate seamlessly.

- **Consistency**: The consistent approach to software development makes it easier to ensure the interoperability, maintainability, and scalability of SWCs.

- **Flexibility**: It promotes a modular and flexible approach to software development, making it easier to adapt to changing requirements and technologies.

In the next section, we will examine the methods used by AUTOSAR to address the limitations of traditional automotive software development. We will investigate how AUTOSAR's common platform, modular design, cost-effective integration, safety and security, interoperability, consistency, and flexibility have transformed the industry by providing a standardized and efficient framework for developing superior automotive software.

How were these goals achieved?

The separation of infrastructure from the application is a fundamental principle of software architecture that enables software developers to focus on their core competencies and expertise in developing **application software** (**ASW**) modules that provide the unique features and functions of their product. They do not need to worry about the underlying hardware or software platform or the implementation details of the BSW and **runtime environment** (**RTE**). Instead, they can focus on their core competencies in developing ASW modules while providing standardized interfaces and services for accessing the BSW.

> **Note**
>
> The BSW provides low-level software services, such as drivers and communication and diagnostic services, that are necessary for the proper functioning of the ECU. It includes standardized modules that are compliant with the AUTOSAR specifications and can be easily integrated and interchanged with other BSW modules from different vendors.

The following figure shows the basic structure of this standardized interface:

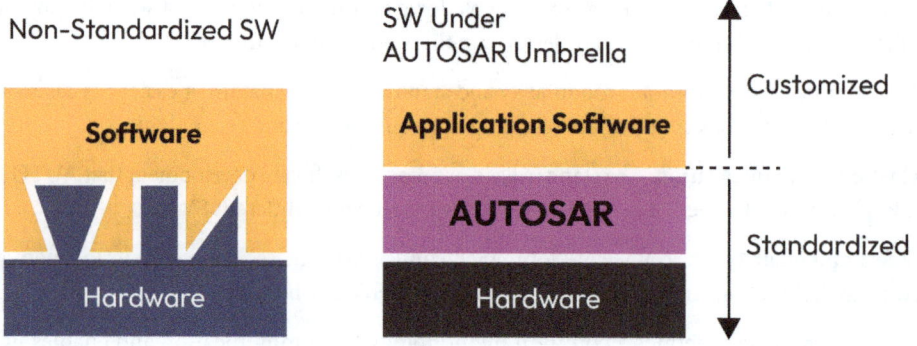

Figure 1.4 – AUTOSAR Standardization

The separation of infrastructure from the application also enables the ECU to be more modular and scalable. Different ASW modules from different vendors can be easily integrated into the ECU, and new ASW modules can be added or removed without affecting the BSW or other ASW modules.

However, AUTOSAR does not provide specific solutions for individual problems or use cases. Instead, it provides a structured way to implement specific functionality in an ECU that can be adapted and customized to meet the needs of different car manufacturers and use cases.

For example, if a car manufacturer wants to implement a specific braking system, they would use the AUTOSAR framework and specifications to develop the software modules that control the braking system. AUTOSAR would provide a standardized way to interact with the microcontroller and peripheral hardware to allow the implementation of the braking functionality. However, it would not provide specific guidance on how to manage the braking system or how to implement specific features or functions. The following case study provides a more detailed example.

Case study – Developing an ECU in the AUTOSAR framework

An automotive supplier is developing a new radar sensor system for autonomous driving applications using AUTOSAR. The challenge is in developing a highly accurate and reliable radar sensor system that can detect objects and provide warning signals to the vehicle's control system as soon and accurately as possible.

The supplier has decided to adopt AUTOSAR, which provides a standardized set of interfaces and services for the development and integration of SWCs. The AUTOSAR BSW layer provides a set of pre-built software modules that abstract the hardware-specific details and provide a unified interface for the ASW layer to access the hardware functions.

By using AUTOSAR, the supplier is able to reduce the development time and cost required to integrate multiple SWCs, as well as reduce the risk of incompatibilities or errors. The standardized interfaces and services provided by AUTOSAR enable the software engineers to focus on developing the application-specific features of the radar sensor system, rather than spending time on developing and integrating basic software services.

For example, the application engineers don't have to worry about how to store data in non-volatile memory, how communication works, or whether to use a **controller area network** (**CAN**) or Ethernet as a medium of transmission.

Instead, they only focused on implementing warning algorithms that would detect objects and provide warning signals to the vehicle's control system.

Now that we've understood the need for AUTOSAR, let's look at its standards in more detail.

Understanding the AUTOSAR standards

The AUTOSAR standards consist of a set of specifications, standards, and guidelines that provide a framework for the development and integration of SWCs in automotive ECUs. The AUTOSAR standards consist of the following components:

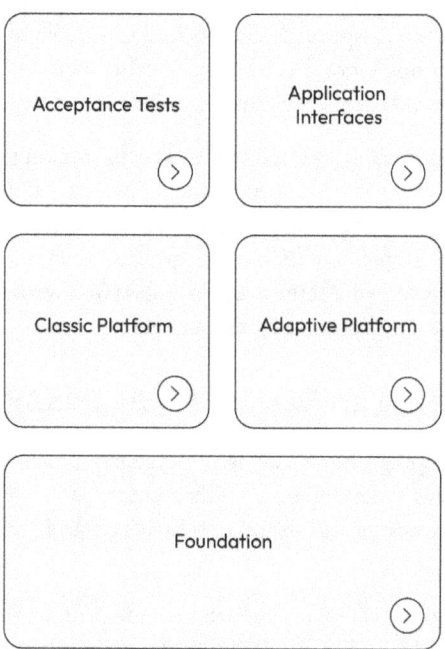

Figure 1.5 – AUTOSAR standards

The following list describes these components in detail:

- **AUTOSAR platform**: AUTOSAR has two main flavors, namely, **Classic Platform** (**CP**) and **Adaptive Platform** (**AP**). It is a common misconception that adaptive AUTOSAR is replacing classic AUTOSAR. Both adaptive AUTOSAR and classic AUTOSAR are complementary software architectures that address different requirements and use cases in the automotive industry:

    - **CP** is intended for conventional embedded automotive systems that have a fixed set of functionalities and tightly integrated hardware and software (static). It features a layered architecture and standardized interfaces that facilitate the consistent and efficient development and integration of SWCs. It is also optimized for resource-constrained environments, making it ideal for ECUs with limited processing power and memory.

        It uses an OSEK/VDX-based **real-time operating system** (**RTOS**), which is highly deterministic and optimized for real-time performance, thus suitable for safety-critical and hard real-time applications such as powertrain and chassis control.

        Communication is based on static configurations using predefined protocols such as CAN, LIN, FlexRay, and Ethernet, and communication relationships are established at design time.

- **AP** is designed for more flexible and dynamic automotive systems, where the hardware and software are decoupled and can be upgraded independently. It provides a modular architecture and supports the use of open source and third-party components, enabling faster innovation and more frequent updates. In contrast to CP, it supports dynamic communication with a **service-oriented architecture** (**SOA**) using Ethernet, where services can be discovered and communicated with during runtime. It uses a POSIX-compliant operating system, typically based on Linux, which provides greater flexibility, suitable for handling complex and resource-intensive tasks.

The two flavors of AUTOSAR reflect the different needs and requirements of the automotive industry, and they are designed to provide a common standard for the development and integration of SWCs in vehicles, regardless of the specific use case or application.

- **Foundation**: **Foundation** (**FO**) in AUTOSAR is designed to ensure interoperability between different AUTOSAR platforms by providing a set of generic artifacts that are shared between both CP and AP. This helps to promote compatibility between different platforms, including those that are not based on AUTOSAR, and ensures that automotive systems and devices can work together seamlessly.

It contains common requirements and technical specifications, for example, E2E Protocol Specification, V2X Specification, Secure Onboard Communication Protocol, and Specification of Intrusion Detection System Protocol, which are all shared between the AUTOSAR platforms (CP and AP).

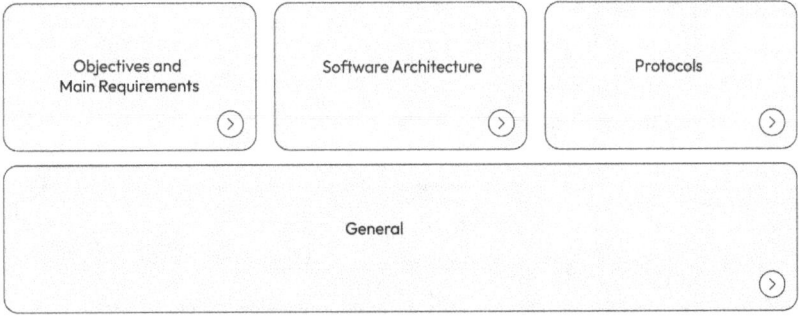

Figure 1.6 – FO within AUTOSAR standard

- **Acceptance tests (ATs)**: AUTOSAR ATs are system tests at the bus level as well as the application level to validate the behavior of an AUTOSAR stack with regard to the ASW components as well as the communication bus:

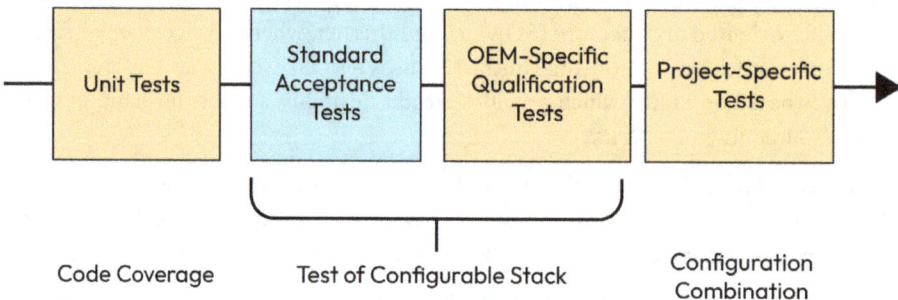

Figure 1.7 – Acceptance test AUTOSAR

The documentation at `https://www.AUTOSAR.org/fileadmin/standards/tests/1-2/AUTOSAR_ATS_CommunicationCAN.pdf` describes AT specifications of communication on a CAN bus. It describes the test case steps along with the test architecture. In *Figure 1.8*, taken from the preceding example use case, this diagram illustrates a test setup where a test bench sends test sequences to an embedded SWC within the **system under test (SUT)** via the CAN bus.

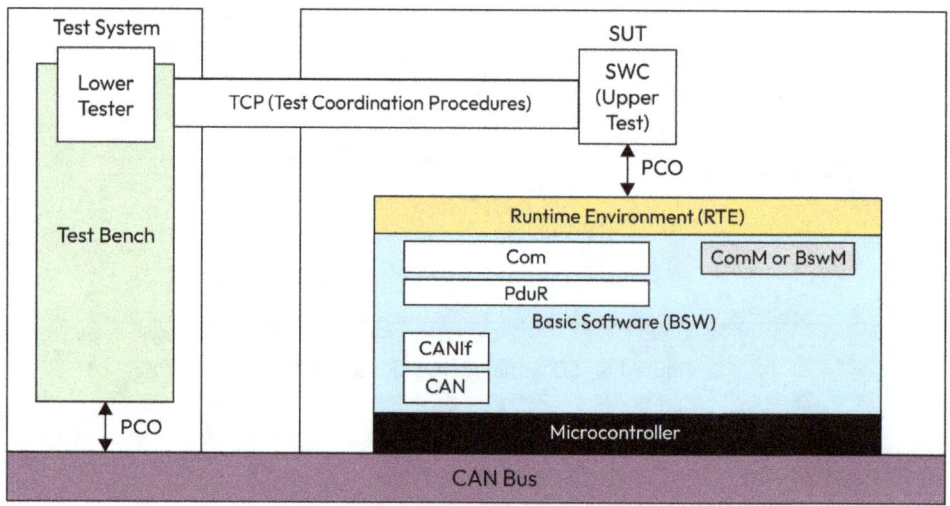

Figure 1.8 – Test architecture description

- **Application interfaces (AIs)**: AUTOSAR has standardized a large set of AIs in terms of syntax and semantics for the different vehicle domains: Body and Comfort; Powertrain; Chassis Control; Occupant and Pedestrian Safety; and HMI, Multimedia and Telematics. For instance, the Chassis Control domain focuses on stability and control, while Powertrain manages propulsion. The Body domain handles comfort and convenience features, autonomous systems enable self-driving capabilities, and **in-vehicle infotainment** (**IVI**) ensures entertainment and connectivity. AUTOSAR defines reference interfaces and SWCs across these domains that the systems can communicate effectively and are interoperable, regardless of the specific hardware or software used by different manufacturers. For example, the Powertrain and Chassis Control systems work together for car stability and performance, while the ADAS relies on inputs from various sensors and systems to make real-time driving decisions.

  An example of an application interface in AUTOSAR is the **adaptive cruise control** (**ACC**) interface. This interface standardizes how the ACC component communicates with other vehicle systems, such as sensors and actuators. It defines how data such as vehicle speed and distance from the car ahead is processed and exchanged, allowing ACC to maintain a safe distance from other vehicles automatically. By using this standardized interface, different ACC implementations can interact with various sensors and actuators across different vehicle models, guaranteeing consistency and interoperability. This is discussed in further detail here: `https://www.autosar.org/fileadmin/standards/R22-11/CP/AUTOSAR_EXP_AIChassis.pdf`

AUTOSAR standardizes the interfaces and SWCs across these domains, ensuring that the systems can communicate effectively and are interoperable, regardless of the specific hardware or software used by different manufacturers. For example, the interaction between the Powertrain and Chassis Control systems is crucial for vehicle stability and performance, while the Autonomous domain relies on inputs from various sensors and systems to make real-time driving decisions.

Software architecture and design

The defined architecture is based on a modular, layered approach that emphasizes the separation of concerns, clear interfaces, and independence of components. AUTOSAR provides guidelines and best practices for software architecture and design, which enable software engineers to develop high-quality, standardized SWCs that can be easily integrated into different automotive systems and devices.

Layers

A **layer** in software architecture refers to a logical grouping of SWCs that share a common set of responsibilities and are designed to work together to perform a specific set of tasks.

It can be seen as a horizontal slice through the software architecture, with each layer providing a specific set of services to the layer above it, as shown in *Figure 1.9*. The layers are typically designed to be modular and loosely coupled so that changes made to one layer do not affect the functionality of the other layers.

The AUTOSAR layered architecture provides the necessary mechanisms for achieving software and hardware independence by dividing the software into three main layers that run on a microcontroller, as shown in the following figure:

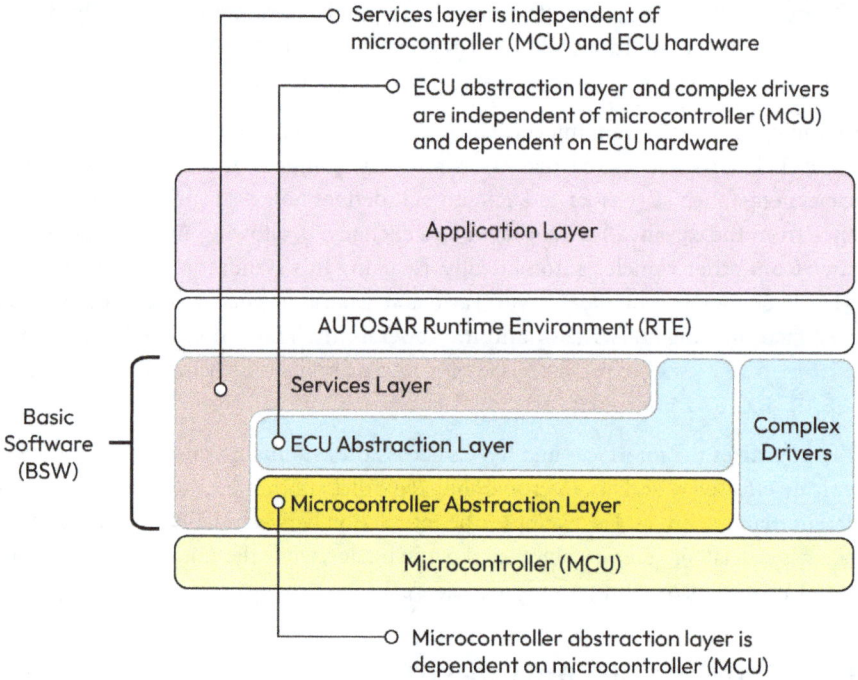

Figure 1.9 – AUTOSAR layered architecture

Let's discuss these layers in further detail:

- **Application layer**: This is where the SWCs that contain the algorithms and functionality of the system are located. This layer is responsible for implementing the high-level behavior of the system and uses the interface of the lower layers to access the hardware resources. Some examples of functions are monitoring the battery charge for a battery charger ECU and setting and viewing the temperature through the **human-machine interface (HMI)**.

- **AUTOSAR RTE**: The RTE serves as both a binding and isolating layer between the ASW and BSW layers. All communication and service usage between these two layers must occur within the RTE. We will dive into the specifics of this topic in greater detail in *Chapter 4*.

- **Basic software**: The BSW layer is divided into three different layers, each providing specific functionality required for the proper functioning of an ECU. These layers are as follows:

  - **Services layer**: This layer offers a variety of services for applications to utilize. It comprises services such as System Services, Memory Services, Crypto Services, and Diagnostic and Communication Services.

  - **ECU abstraction layer**: This layer delivers ECU-related abstractions, including I/O Hardware Abstraction, Onboard Device Abstraction as an external watchdog, Memory Hardware Abstraction, and Crypto Hardware Abstraction, to enable hardware independence for applications.

  - **Microcontroller abstraction layer** (**MCAL**): This layer provides a driver implementation for the MCU in use, enabling communication between the BSW layers above and the microcontroller hardware peripherals.

Further exploration of the layers will be conducted in *Chapter 2*.

Stacks

A **stack** refers to a collection of software layers or components that work together to provide a specific functionality or service. Each layer in the stack represents a distinct set of functions and services and is responsible for performing a specific set of tasks. The layers communicate with each other through well-defined interfaces, and the data is passed between the layers in a hierarchical manner, with each layer providing services to the layer above it. The architecture is shown in the following figure:

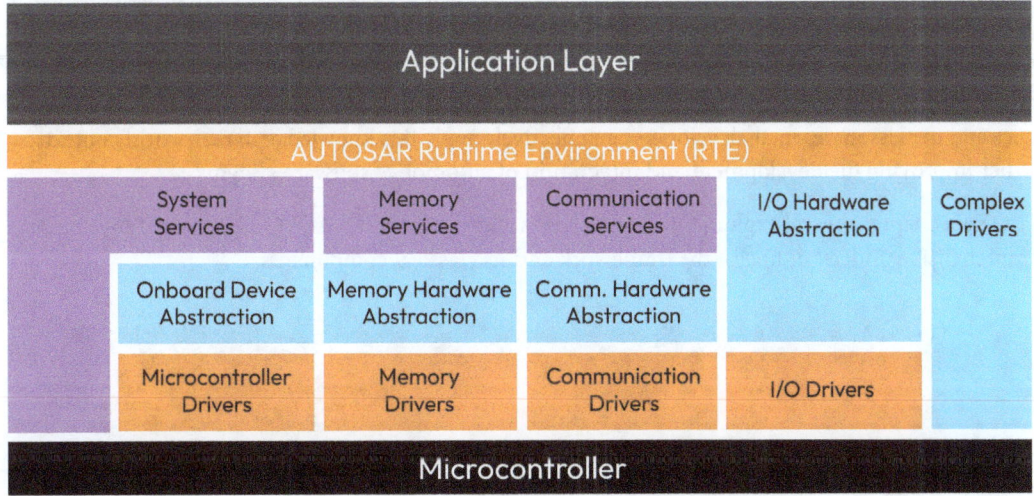

Figure 1.10 – AUTOSAR stacks

Some of the main AUTOSAR stacks that we will dive into in our journey are the following:

- Memory Stack
- Communication (CAN, Ethernet, LIN, and FlexRay)
- Diagnostic
- IO Hardware Abstraction
- Security

Summary

Throughout this chapter, we have traced the evolution of the automotive industry and identified the challenges that traditional software development faces. These challenges served as key driving forces behind the creation of AUTOSAR. We discussed that the AUTOSAR framework offers a standardized method for software development and integration within the automotive sector. This enables software developers to concentrate on their core competencies while developing application-specific features and taking advantage of standardized interfaces and services to access the basic software layer. Furthermore, the framework fosters interoperability, scalability, and the reuse of SWCs across various car manufacturers, ultimately reducing development time and costs while enhancing software quality and reliability.

We learned that AUTOSAR is comprised of specifications and guidelines that establish a framework for the development and integration of SWCs in automotive ECUs. The framework employs a layered approach with modular and loosely coupled layers, achieving ASW and hardware independence. Additionally, AUTOSAR provides a set of stacks that collaborate to deliver ECU-specific functionalities and services. The insights gained in this chapter are significant because they allow for a better understanding of the real-world applications of AUTOSAR. By comprehending the relevance of these topics and the lessons learned, developers can effectively apply this knowledge in real-world contexts, further improving the development and integration of automotive software systems.

In the next chapter, we will explore the various layers and stacks within AUTOSAR, providing a more comprehensive understanding of the framework and its classical components.

Questions

Please answer the following questions to evaluate your overall understanding of this chapter:

1. What does AUTOSAR stand for?
2. Who developed the AUTOSAR standard?
3. What are the main goals of the AUTOSAR framework?
4. How many layers does classical AUTOSAR have?
5. List four of the AUTOSAR stacks.
6. Is adaptive AUTOSAR replacing classical AUTOSAR, and if so, why?
7. What are the key differences between classical and adaptive AUTOSAR?

Get This Book's PDF Version and Exclusive Extras

Scan the QR code (or go to `packtpub.com/unlock`). Search for this book by name, confirm the edition, and then follow the steps on the page.

Note: Keep your invoice handy. Purchases made directly from Packt don't require one.

2
Introducing the AUTOSAR Software Layers

By now, you have learned about AUTOSAR, its challenges, and its development by a group of collaborators who joined forces to create it. But what is AUTOSAR exactly?

Consider AUTOSAR as a collection of modular elements that create a base platform upon which automotive applications can be built. Similar to how building blocks can be used to construct various structures, AUTOSAR offers a range of standardized modules and **software components** (**SWCs**) that can be combined to create diverse automotive systems.

> **What is an SWC?**
> In AUTOSAR, an SWC is a modular, reusable, and self-contained piece of functionality that encapsulates a specific part of a software application and represents various types of functionalities, such as sensor data processing, actuator control, or complex algorithms. AUTOSAR offers a standardized way of defining these components besides offering standardized interfaces to the underlying service components.

AUTOSAR can be likened to a universal language for automotive software development. In the same way that people from different countries can communicate using a shared language, SWCs created with AUTOSAR can interact with one another through standardized interfaces and protocols, irrespective of the underlying hardware or software platform.

In this chapter, we will attempt to understand this in depth by covering the following main topics:

- Understanding layered software architecture in AUTOSAR
- Application layer
- Runtime environment
- Basic software layer

- Interfaces
- Case study – Developing a BMS ECU

Understanding layered software architecture in AUTOSAR

We can visualize the layered software architecture in AUTOSAR as a multi-story building, where each floor serves a distinct purpose and has its own set of responsibilities. In this analogy, the floors represent the different layers of the AUTOSAR software architecture, and the rooms on each floor represent the modules within each layer:

- **Basic software (BSW) layer (basement)**: This floor represents the foundation of the building, providing essential services such as plumbing, electrical systems, and heating. Similarly, this layer contains modules that manage low-level tasks such as communication, diagnostic services, memory management, and **operating system (OS)** functionalities.
- **Runtime environment (ground floor)**: This floor serves as the hallway that connects the ground floor to the upper floor, facilitating smooth movement between them. The **runtime environment (RTE)** in AUTOSAR facilitates communication between SWCs in the application layer and the BSW layer, providing an abstraction for the underlying hardware.
- **Application layer (upper floor)**: This floor contains the living spaces, such as bedrooms, kitchens, and living rooms, which cater to the specific needs of the occupants. In the same way, the application layer houses application-specific SWCs that define the functionality and behavior of the automotive system.

In this multi-story building analogy, as shown in *Figure 2.1*, each floor serves a unique purpose and relies on the services provided by the floors below it:

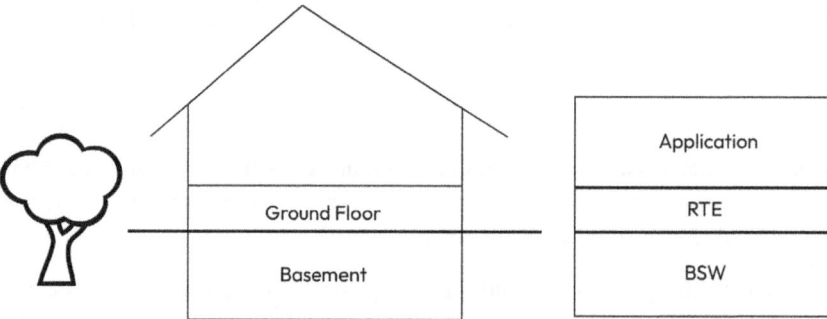

Figure 2.1 – AUTOSAR analogy

The building's occupants (the end users) primarily interact with the upper floor (the application layer), which is specifically designed to meet their needs. Similarly, the automotive system's functionality is primarily dictated by the application layer, which utilizes the lower layers to achieve its goals.

This analogy helps illustrate the distinct roles and dependencies of each layer within the AUTOSAR software architecture, highlighting the importance of abstraction, and separation of concerns in managing the complexity of automotive systems. That said, let's discuss and investigate the AUTOSAR layers shown in *Figure 2.2*. Here is what the general structure looks like:

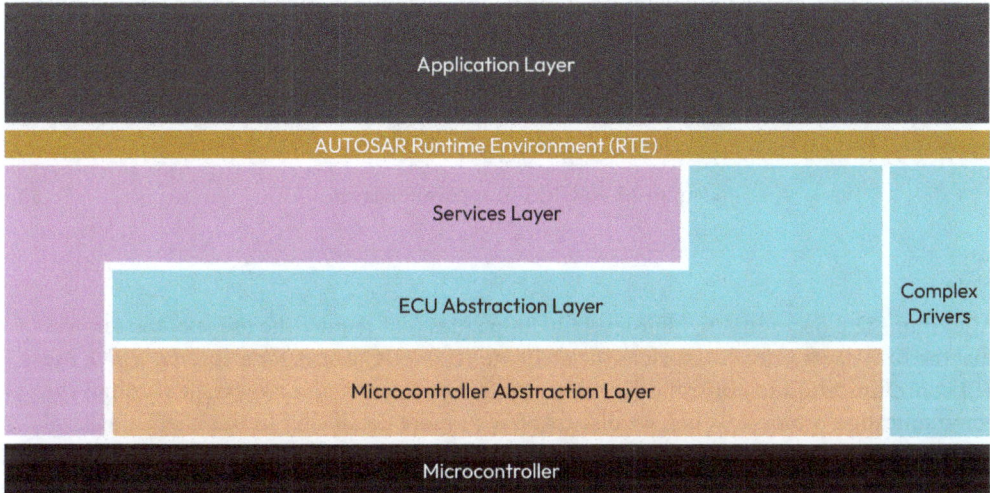

Figure 2.2 – AUTOSAR layers

We will adopt a top-down approach in this chapter, discussing the functionality of each layer in greater detail, and we will conclude our discussion with a case study discussing the development of a **battery management system (BMS)** ECU. Let's look at the application layer first.

Application layer

Within AUTOSAR, the **application layer** represents the highest level of the software architecture and is tasked with executing the system's high-level functions.

This layer consists of SWCs designed to carry out specific tasks or operations needed by the automotive system. These components can encompass control algorithms, signal processing, monitoring and diagnostic functions, user interfaces, and other application-specific logic. The SWCs communicate with each other and the lower layers via well-defined interfaces, promoting modularity, reusability, and maintainability. The AUTOSAR software architecture defines the application layer, and its components are engineered to be independent of the underlying hardware and BSW layer. This independence allows for increased flexibility and reusability of the SWCs across various automotive systems. It can be visualized as shown in the following figure:

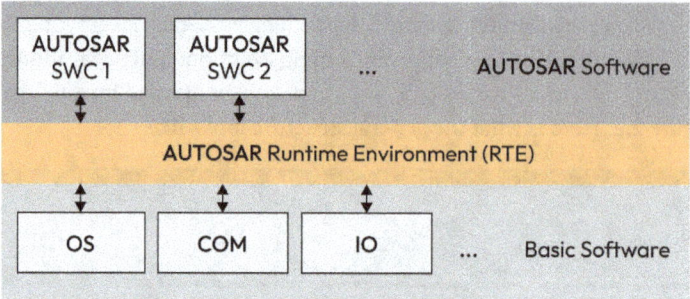

Figure 2.3 – AUTOSAR application layer

> **Note**
> Algorithms and decisions are high-level functionalities that require the usage of the services that the BSW layer provides, such as the ability to read ADC values for a specific GPIO, use SPI communication to control other chips on the ECU, and send a message through the communication protocol. When we discuss BSW in more detail later in this chapter, we will get even more familiar with the services BSW provides to the application.

For example, a radar sensor ECU is responsible for processing and interpreting data from RF chips, which are typically used for **advanced driver assistance systems** (**ADAS**) and autonomous driving features. Blind spot detection is a feature that uses radar sensors to detect when a vehicle is in your blind spot and warns you with an audible or visual alert. The application layer, in this case, will consist of several SWCs, each responsible for a specific part of the functionality. For instance, there may be a component that reads the raw data from the radar sensors; another component tracks and classifies the objects; another component processes the data to detect potential hazards; and another component generates visual or auditory alerts for the driver.

Would the application layer care about the underlying communication protocols? No. If the ECU sends the detected objects or warning signals over a communication bus, the SWCs send this data to the BSW layer and its service modules pack this data into the configured communication bus (**controller area network** (**CAN**), Ethernet, or FlexRay).

> **Note**
> Although AUTOSAR doesn't prescribe a specific programming language, most of its SWCs are written in C. This is because C is a widely recognized and well-established language that is highly compatible with the development of low-level system software, such as device drivers and communication protocols.

In the upcoming section, we will delve deeper into the RTE layer, which serves as a bridge connecting the application layer and the BSW services.

Runtime environment

The RTE plays a critical role in the AUTOSAR architecture; it is the middleware layer facilitating communication between SWCs in the application layer and the BSW layer. The RTE serves as an abstraction layer, enabling the exchange of data and signals between SWCs without needing to know the underlying hardware or software details.

This abstraction enables SWCs to be designed, developed, and tested independently of each other and the target hardware platform. In addition, the RTE provides a range of services, such as data marshaling, data transformation, and event-driven communication, further enhancing the efficiency and robustness of the overall system.

Before discussing RTE further, we need to get familiar with the **virtual functional bus (VFB)** concept so that we can better understand the RTE.

Understanding the VFB concept

The VFB is an abstract concept in AUTOSAR that represents the communication infrastructure for SWCs in the application layer. It defines a set of standardized interfaces and protocols for data exchange between SWCs, allowing them to communicate without knowledge of the underlying hardware or software details.

But what is this even supposed to mean?

The VFB is necessary during the *configuration step* where the process of defining and setting up the architecture of an AUTOSAR system takes place. This includes defining the SWCs, their interfaces, and the communication between these components.

This is also when architects decide on the functionality needed in a system, thus dividing these functionalities into different SWCs:

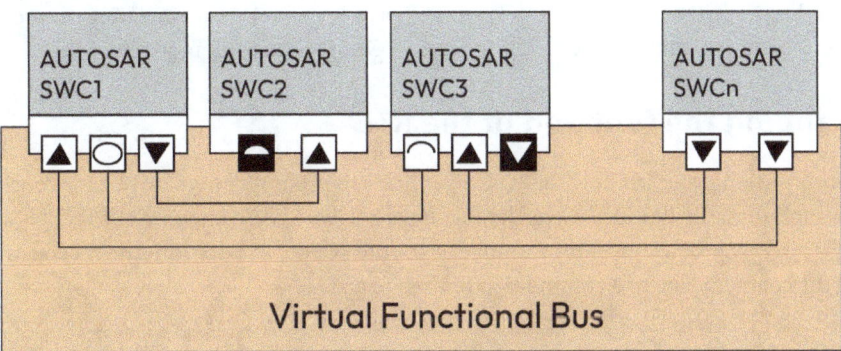

Figure 2.4 – VFB concept

Each piece of software is decomposed into multiple SWCs, each with its own dedicated functionality. When an SWC needs to exchange data with another SWC, this is done using signals or even **remote procedure calls** (**RPCs**). Once the software architecture has been completed, the SWCs are distributed across one or more physical ECUs. These ECUs are responsible for implementing the functionalities as well as facilitating communication between the components:

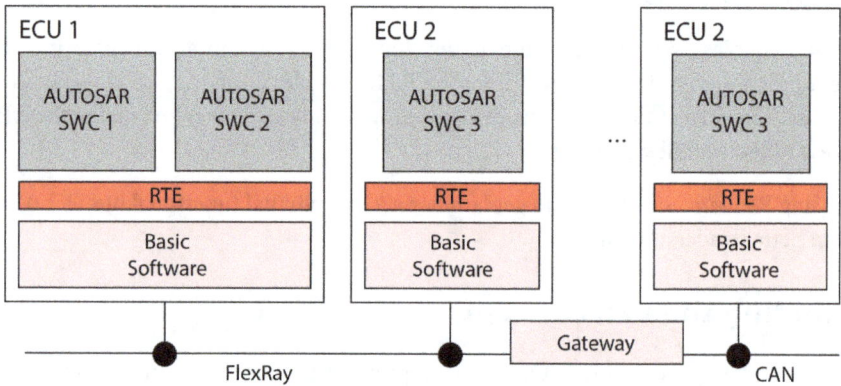

Figure 2.5 – Allocating SWCs to ECUs

So, if an SWC, such as *SWC 1* and *SWC 2*, is within the same ECU, the communication would be intra-ECU communication mechanisms where the communication would remain within the ECU, which would then be realized by the RTE.

Yet, if SWCs are located in two different ECUs, such as *SWC 1* and *SWC 3*, the communication between them would be realized through a communication medium, such as CAN, FlexRay, or Ethernet. The process of how signals are packed within the BSW layer will be discussed in deeper detail in *Chapter 5*.

For now, consider the RTE as a bridge or glue layer, allowing SWCs to share data with one another while also enabling interaction with the BSW layer. This plays a vital role in maintaining the modularity, reusability, and interoperability of the SWCs in an AUTOSAR-based automotive system.

Understanding the function of the RTE

You are probably asking yourself how the RTE layer helps to promote modularity. Let's consider the following example to clarify this idea. Consider that there are two SWC components; *SWC 1* is a sender and *SWC 2* is a receiver of a signal, for example, a counter value. In addition, the receiver module would send the counter value to a communication bus (e.g., CAN).

> Note
> The following example is just to demonstrate how the RTE layer acts as the glue layer between SWCs and the BSW service layer.

The following example defines a function within *SWC 1* called `SenderCounterFunction`. This function increments a static variable called `counter` and writes its updated value using a function called `Rte_Write_SenderInterface_SenderOutput`:

```
#include "Rte_Sender.h"
void SenderCounterFunction(void) {
    static UInt32 counter = 0U;
    counter++; /*just an example */
    Rte_Write_SenderInterface_SenderOutput(counter);
}
```

SWC 2 defines another function called `ReceiverCounterFunction`. Inside the function, a local variable, `receivedData`, of type `UInt32` is declared. The function then reads a value and stores it in the `receivedData` variable.

The code then checks whether `receivedData` is odd or even. If it's odd, the function sends the value using a function called `Rte_Write_ReceiverModule_CounterSig`:

```
#include "Rte_Receiver.h"
void ReceiverCounterFunction(void) {
    UInt32 receivedData;
    Rte_Read_ReceiverInterface_ReceiverInput(&receivedData);
    // Process received data, e.g., control an actuator based on the input
    // In this example only odd counter value is sent
    if (receivedData % 2) {
        Rte_Write_ReciverModule_CounterSig(receivedData)
    }
}
```

The following code demonstrates a possible implementation of an RTE layer, which acts as an intermediary between SWCs or applications. The RTE layer serves as the *glue* that enables signals to be communicated between the different SWCs. To achieve this, the code defines a buffer that acts as a temporary storage location for data and provides setter and getter functions that can be used to read from or write to the buffer. Overall, this code provides an example of how the RTE layer can be used to facilitate communication between SWCs:

```
UInt32 Rte_buffer_Data=0U;
FUNC (Std_ReturnType, RTE_CODE) Rte_Write_SenderInterface_
SenderOutput(UInt32 data)
{
  Std_ReturnType ret = RTE_E_OK;
    Rte_buffer_Data = data;
    return ret;
}
```

```
FUNC(Std_ReturnType, RTE_CODE) Rte_Read_ReceiverInterface_
ReceiverInput (UInt32 *data)
{
   Std_ReturnType ret = RTE_E_OK;
    *data = Rte_buffer_Data;
   return ret;
}
FUNC (Std_ReturnType, RTE_CODE) Rte_Write_ReciverModule_
CounterSig(UInt32 data)
{
   Std_ReturnType ret = RTE_E_OK;
   ret |= Com_SendSignal(ComConf_ComSignal_CounterSig_Tx,&(data));
   return ret;
}
```

As observed in this example, two SWCs communicate with each other by exchanging data. The RTE manages the buffering and data handling process. The two SWCs function separately, where *SWC 1* sends the data to the RTE, and the RTE stores the data internally. Whenever the other SWC needs the data, it requests it from the RTE. Moreover, when the data is being transmitted to the communication bus, the COM service of the BSW layer is employed. Please take a further look at *Figure 2.6* where this process is shown as a sequence diagram. In *Chapter 7*, we will explore the complete pathway from the application layer to the physical communication bus.

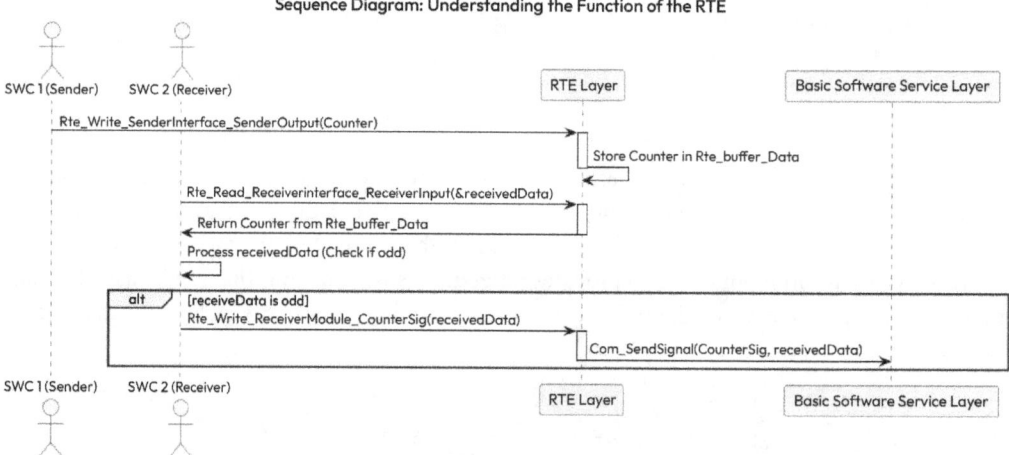

Figure 2.6 – RTE layer as a mediator between SWCs

At this point, we have acquired an understanding of the mechanics and operations of both the application and RTE layers. To gain a more in-depth understanding, let us explore the BSW layer.

Basic software layer

At the lowest level of a smartphone, you have the hardware components, such as the processor, memory, and sensors. These components are managed by device drivers and other low-level software modules that allow them to communicate with the higher-level SWCs, such as your beloved WhatsApp or TikTok applications. In AUTOSAR, the BSW layer provides this low-level functionality, such as memory management, communication protocols, and diagnostic services, to the upper application layer, as shown in the following figure.

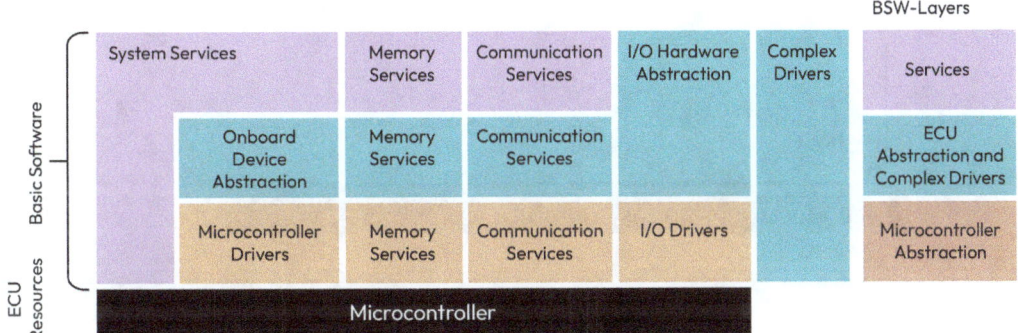

Figure 2.7 – BSW layers

Let's zoom into the architecture and the different sublayers of the BSW, which is composed of three additional sublayers: the service layer, ECU abstraction layer, and microcontroller layer.

Service layer

The topmost layer is the **service layer**, which provides various services for the application components to use. It is the primary interface between application components and the BSW functionality. It is the only layer to which the application has access, and it enables the application components to access and utilize a range of functionalities without having to deal directly with the low-level hardware or SWCs. Let's highlight these components in the BSW diagram:

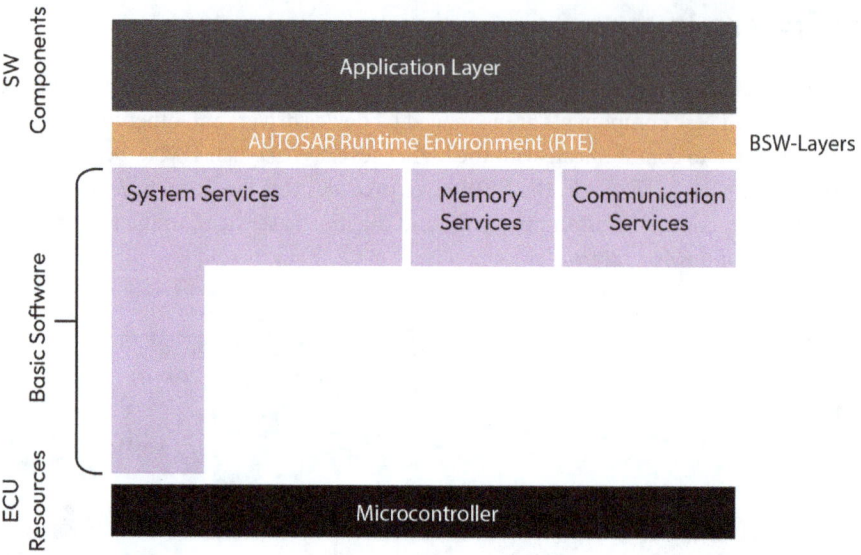

Figure 2.8 – BSW Service Layer

Examples of services that the BSW provides to the application are the following:

- **Communication services**: These services handle the communication between different ECUs and between the ECU and external devices, using standardized protocols such as CAN, LIN, FlexRay, and Ethernet. By providing generic communication services, the BSW layer allows for the seamless integration of various components in a distributed automotive system.

    If your application needs to send CAN messages, then it utilizes the COM APIs to trigger the sending of messages. You will learn more about this in the *Chapter 7*.

- **Memory services**: The BSW layer provides standardized memory services, such as EEPROM Abstraction and Flash EEPROM Emulation, enabling the application layer to access and manage nonvolatile memory without dealing with hardware-specific details.

    For example, the ECU under development could need to permanently store data such as calibration values, configuration parameters, and other important settings.

- **Diagnostic services**: Diagnostic services, such as **diagnostic communication manager (DCM)** and **diagnostic event manager (DEM)**, enable the application layer to perform diagnostics and report errors. These services provide a standardized approach to diagnostics, which simplifies development and maintenance. More about this in *Chapter 10*, when we touch briefly on the diagnostics modules.

- **System services**: These services manage essential system functions, such as **ECU state manager** (**ECUM**), the OS, and **mode management**. By offering generic system services, the BSW layer allows the application layer to focus on high-level functionality without worrying about low-level system management tasks.

ECU abstraction layer

The **ECU abstraction layer** in AUTOSAR serves as a universal connector, allowing diverse software AUTOSAR components to interact with and manage distinct hardware devices without the need for knowledge about each device's specific details. This layer establishes a standardized interface between the BSW and the hardware, ensuring compatibility and seamless operation across various ECUs, independent of the particular manufacturer or model of the microcontroller, the ECU configuration, or the specifics of whether certain hardware peripherals are integrated into the microcontroller or exist as separate chips. By abstracting the disparities between ECUs, the ECU abstraction layer enables the application layer to be developed without being tied to a specific hardware implementation or specific layout of the ECU:

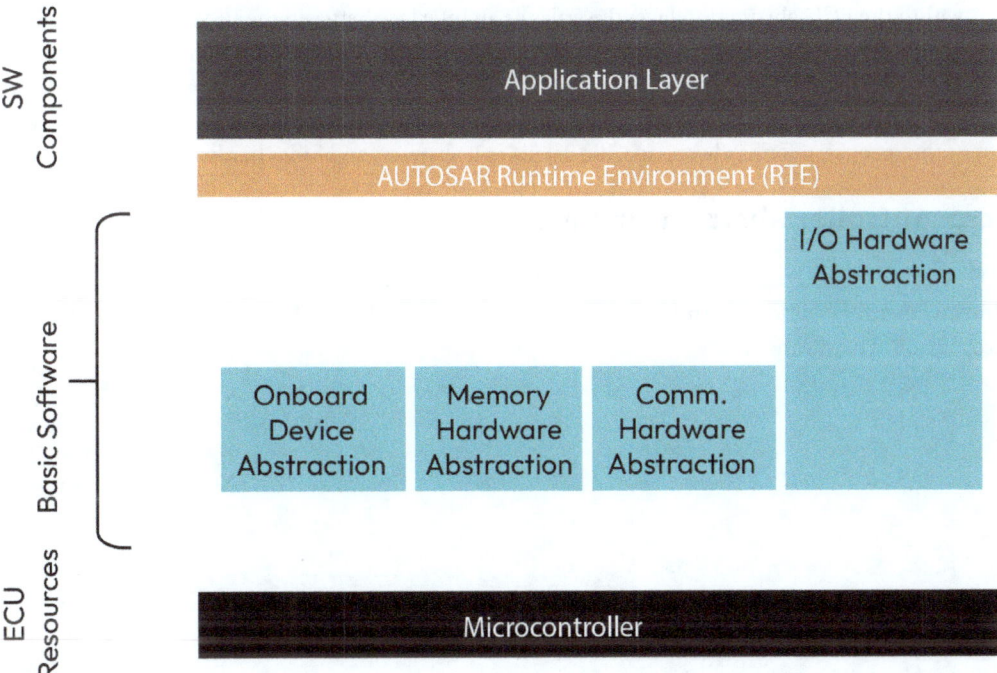

Figure 2.9 – ECU abstraction layer

In a complex automotive system, an ECU might contain multiple CAN devices, both internal and external. These devices are responsible for communication with other ECUs and systems within the vehicle. The ECU abstraction layer plays an important role in managing the interactions with these various CAN devices while maintaining the application's independence from the underlying ECU layout. The ECU provides the following features in this case:

- **Unified interface**: The ECU abstraction layer provides a unified interface to the application for interacting with all the CAN devices in the system, regardless of whether they are internal or external. This abstraction hides the complexities and differences between the various CAN devices, allowing the application code to remain independent of the specific ECU layout.
- **Adaptability**: The ECU abstraction layer's ability to manage the interactions with multiple CAN devices allows the application to be more adaptable to changes in the ECU layout. For example, if a new CAN device is added or an existing one is replaced, the ECU abstraction layer can be updated to accommodate these changes without requiring modifications to the application code. This makes the system more scalable and easier to maintain.

Another example would be **Input/Output Hardware Abstraction**, positioned below the RTE within the ECU abstraction layer. It serves as an interface between the BSW modules and the application software. It also provides a consistent interface for interacting with various hardware components, such as **digital input/output** (**DIO**), **analog-to-digital converters** (**ADCs**), and **pulse-width modulation** (**PWM**) peripherals, allowing the application layer to function without concern for hardware-specific details as it abstracts and hides hardware-specific details.

Microcontroller abstraction layer

In AUTOSAR, the **microcontroller abstraction layer** (**MCAL**) is the lowest layer in the software architecture and provides a standardized interface between the BSW layer and the underlying microcontroller hardware:

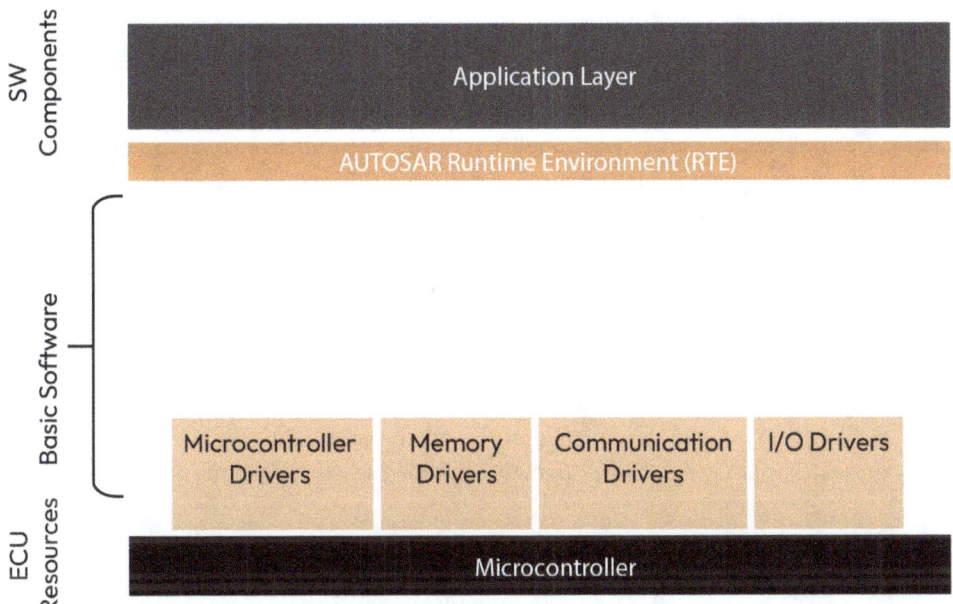

Figure 2.10 – MCAL

The primary purpose of the MCAL is to abstract the hardware-specific details. It offers a set of software modules and device drivers that are tailored to the specific microcontroller used in the system. These modules and drivers provide access to the microcontroller's peripherals. These modules include the DIO ports, ADCs, timers, communication interfaces (CAN, LIN, FlexRay, and Ethernet), drivers and memory management (**Flash Driver** or **FLS**), and Crypto Driver.

Imagine an automotive application that needs to control the speed of a motor using a PWM signal. Without the MCAL, the application developer would need to write code specific to the hardware being used (e.g., a particular microcontroller and PWM module). This could make the application difficult to maintain and port to other hardware platforms. To illustrate this more clearly, in the following example, we will demonstrate how to control an LED using the ECU abstraction layer in AUTOSAR. The goal is to turn the LED on and off periodically, utilizing DIO.

Now, let's look at the code used to toggle the LED:

```
void ToggleHeartBeatLED(){
    static boolean ledState = FALSE;
     // Toggle the LED state
    ledState = !ledState;
    // Write the new state to the LED pin
    Dio_WriteChannel(LED_PIN, ledState);
}
```

Examining the provided code, a unified API, `Dio_WriteChannel`, is employed to access PINs without having to worry about the specific microcontroller implementation. This allows for seamless integration with microcontrollers from various vendors, such as Infineon, NXP, ST, or any other manufacturer.

If we cannot use an abstraction layer to accomplish the task, we would need to access the peripheral registers directly to implement it. However, this approach would tightly couple our application or implementation to a specific family of microcontrollers. This could hinder modularity and make the code less usable across different microcontroller platforms.

Having discussed the BSW layers and their respective functions, which consist of multiple modules to achieve ECU features, it is essential to explore the potential for introducing new features or integrating legacy code into the AUTOSAR system. AUTOSAR strives to uphold a flexible platform while enabling legacy code integration. In response to this requirement, **complex device drivers** (**CDDs**) have been implemented, filling the void and empowering users to add new capabilities to any of the BSW layers not initially present within the AUTOSAR framework. Let's discuss these in the next section.

Complex drivers

In AUTOSAR, a CDD is a specialized software module designed to manage and control complex hardware devices or subsystems that are not covered by other layers. CDDs were created to address the needs of specific automotive systems that require unique functionality or hardware control that goes beyond the specification.

The following figure shows that complex drivers represent a cross-section of all the BSW sublayers:

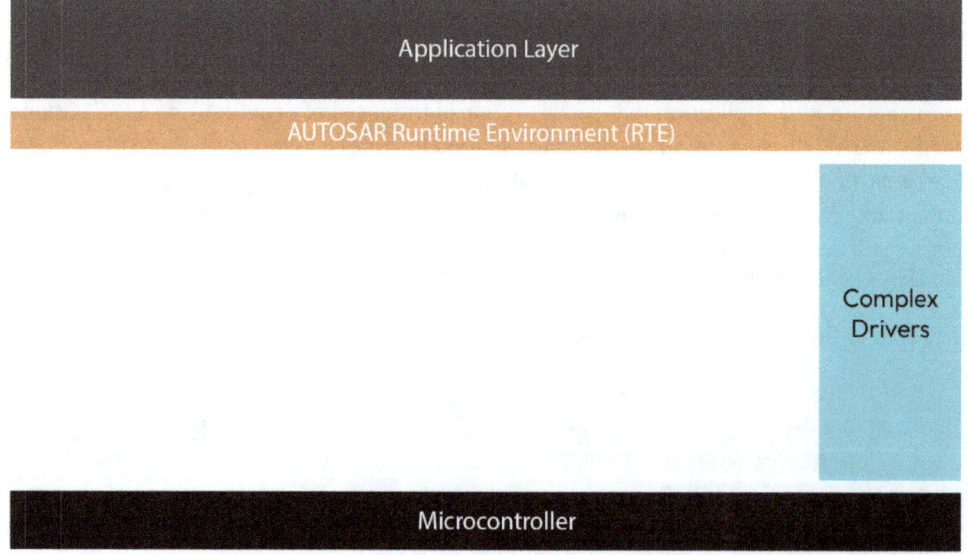

Figure 2.11 – CDDs

Imagine an automotive application utilizing a custom touch-sensing interface to control various in-car features, such as the infotainment system, climate control, and lighting. The touch-sensing interface, developed by a third-party vendor, is not directly compatible with standard AUTOSAR modules. As a result, a CDD is developed to bridge this gap. The CDD may communicate with the sensor through the SPI, utilizing the AUTOSAR SPI module, and subsequently provide the touch information to the application via the RTE, to help the application SWC to take proper actions.

So far, we have discussed the AUTOSAR layered architecture and the components of the BSW layer. To complete the picture, let's discuss the different interfaces between the layers, identifying which are standard and which can be customized by the user.

Interfaces

Interfaces enable communication and interaction between various SWCs and layers. As depicted in the following figure, the different AUTOSAR interfaces can be observed between the application and RTE, between the RTE and BSW, and among the BSW modules themselves:

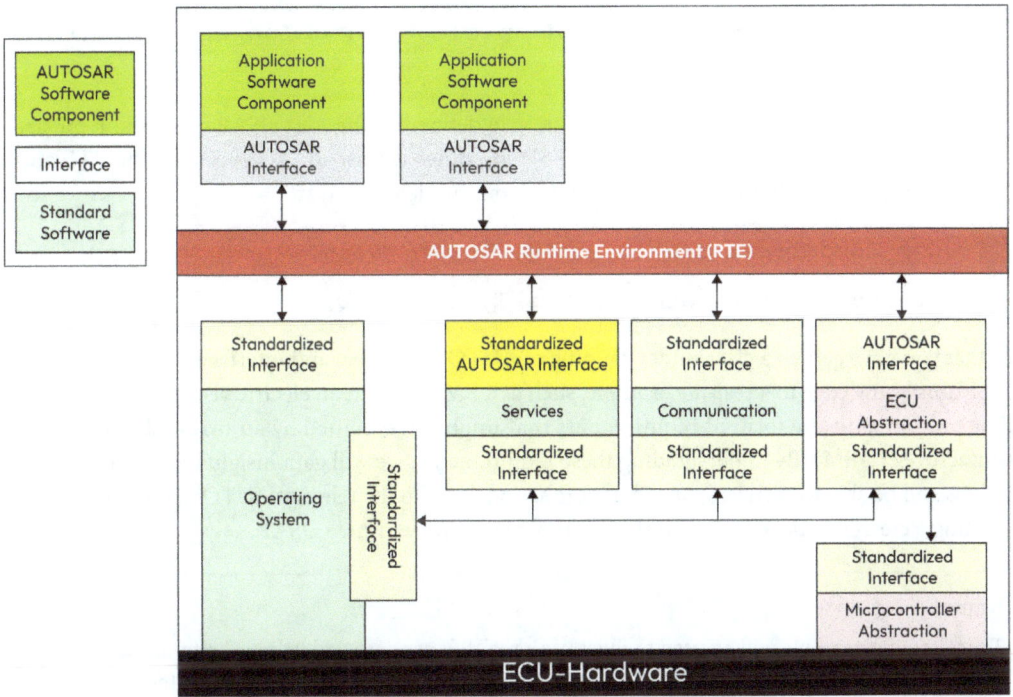

Figure 2.12 – Component view of the AUTOSAR layered software architecture

As observed in the preceding figure, there are three main types of interfaces provided in AUTOSAR:

- **Standardized interfaces**: A standardized AUTOSAR interface is a generic interface originating from the ports of an SWC. Provided by the RTE, these interfaces serve as communication channels between SWCs or between an SWC and the ECU firmware (IoHwAb or complex drivers). Through these interfaces, an SWC can, for instance, read input values and write output values.

  The following interfaces are a part of the standardized AUTOSAR interface structure, which can be utilized by various SWCs through `Rte_read` and `Rte_call`:

  - `Std_ReturnType Rte_Read_<portname>_<interface>(<data>)`
  - `Std_ReturnType Rte_call_<portname>_<operation>(<data>)`

- **Standardized AUTOSAR interfaces**: The syntax and semantics of these interfaces are predefined by the AUTOSAR standard. These interfaces are utilized by SWCs to access AUTOSAR services provided by BSW modules within the service layer, such as ECUM or DEM.

  For example, `Det_ReportError` is an interface used by all BSW modules to report an error:

  - `Det_ReportError(uint16 ModuleId, uint8 InstanceId, uint8 ApiId, uint8 ErrorId);`

- **AUTOSAR interfaces**: These interfaces are also defined by the AUTOSAR consortium but are specific to the interaction between the BSW modules within an ECU, between the RTE and the system OS, or between the RTE and the communication layer:

  - `Com_SendSignal(uint16 SigId,uint8* value)`
  - `StatusType ActivateTask(TaskType TaskID)`

Now that you have gained a clear understanding of AUTOSAR layers and interfaces, it is a good idea to apply this knowledge to a real-life example, such as a BMS ECU in an electric vehicle. This example will be presented in the form of requirements that might be specified by an **original equipment manufacturer (OEM)**. By understanding these requirements, we will gain insight into the necessary functions and be able to correlate the different functionalities to the relevant AUTOSAR layer, thereby facilitating a clearer understanding of the layers for those new to AUTOSAR.

> **Note**
> The following example will be used to illustrate a real-life use case and help us better understand the various functionalities of the different AUTOSAR layers. It should be noted that the example is simplified for the purpose of conveying a message and should not be considered a design reference.

Case study – Developing a BMS ECU

A BMS is responsible for monitoring and managing the battery pack in an electric vehicle. Its primary functions include measuring battery voltage, current, and temperature, estimating the state of charge and state of health, ensuring safe charging and discharging, and providing diagnostic information to other ECUs. The following figure illustrates this workflow:

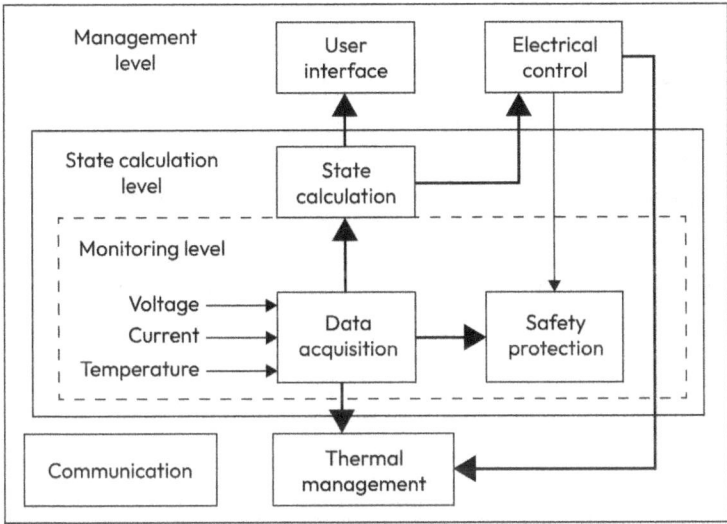

Figure 2.13 – BMS block diagram

The preceding figure shows a proposed system design for a BMS. The following are the ECU requirements in the context of this case study:

1. **State of charge estimation** [REQ_1]: The ECU estimates the remaining battery capacity using algorithms based on battery voltage, current, temperature, and historical data.
2. **Cell balancing** [REQ_2]: The ECU monitors and controls the charging/discharging process of individual cells to ensure optimal performance and extend battery life.
3. **Thermal management** [REQ_3]: Monitor the battery temperature and control the cooling/heating systems to maintain the battery within its optimal temperature range.
4. **Fault detection and protection** [REQ_4]: Detect and manage faults, such as overvoltage, undervoltage, overcurrent, or overtemperature, and take appropriate actions.
5. **Fault storage** [REQ_5]: Faults are stored in a nonvolatile memory.
6. **Diagnostic services** [REQ_6]: Provide diagnostic functionalities (e.g., **unified diagnostic services**, or **UDS**) to monitor the BMS's health and support troubleshooting and maintenance activities.

7. **Communication services** [*REQ_7*]: Facilitate interaction between the BMS and other vehicle systems or external devices (e.g., charging stations) using standardized interfaces and protocols.
8. **Communication** [*REQ_8*]: BMS will send two cyclic state messages (`Msg_A`, `Msg_B`) on CAN.
9. **Memory management** [*REQ_9*]: Handle nonvolatile memory access, such as reading and storing calibration data and fault codes in eFlash memory.
10. **OS** [*REQ_10*]: Manage the scheduling and execution of software tasks, interrupts, and resource allocation within the ECU.
11. **I/O Hardware Abstraction** [*REQ_11*]: Provide an interface for accessing and controlling the battery-related hardware components (e.g., ADC, DAC, or GPIO) and peripheral devices (e.g., temperature sensors, contactors, or cooling system components).
12. **Network management** [*REQ_12*]: Manage the ECU's participation in the vehicle network, including startup and shutdown procedures, sleep/wake-up modes, and error handling.

Now let's discuss how some of these requirements would be implemented and to which layers they would belong.

Application layer

Some of the requirements that could be implemented in the scope of an application layer are *REQ_1* and *REQ_2* – these two requirements state that the SW shall perform battery cell balancing and estimation for the remaining charge, the SW would achieve this functionality by implementing specific algorithms (which are beyond the scope of this book), which makes use of the battery voltage and current.

RTE

As we just discussed, the RTE only acts as a bridge or glue layer; all communication that takes place between the SWC or BSW would be handled through the RTE.

For example, say an SWC called `BattEst` is responsible for implementing the algorithm that estimates battery capacity and transmits the charge percentage. In this scenario, the application utilizes the communication service from the BSW, with the RTE serving as an intermediary between the SWC and the BSW service provider.

BSW – Service layer

All the services the software needs are provided through the service layer, for example, diagnostics services (*REQ_6*) where the software provides the BMS health state, or memory services (*REQ_5*), where the software needs a mechanism to store data in nonvolatile memory.

BSW – ECU abstraction layer

Any service used by the application, sending a message or storing a block in nonvolatile memory, has an impact on the ECU abstraction layer configuration, for example, *REQ_5* storing a data block in nonvolatile memory.

BSW – MCAL

In this stage, all the low-level drivers are configured. For example, according to *REQ_8*, communication over CAN is necessary; thus, the CAN driver must be configured in the MCAL to allow transmission to the physical CAN.

Probably, at this stage, it may be a good idea to showcase an abstracted diagram for interaction or communication between the various layers in the case of message sending and reception. *Figure 2.14* illustrates the sequence of events involved in the transmission of a signal (**Msg_A**) from a software component (**SWCA**) to the CAN bus and the subsequent reception of a different signal (**Msg_B**) by another software component (**SWCB**) through the different AUTOSAR layers.

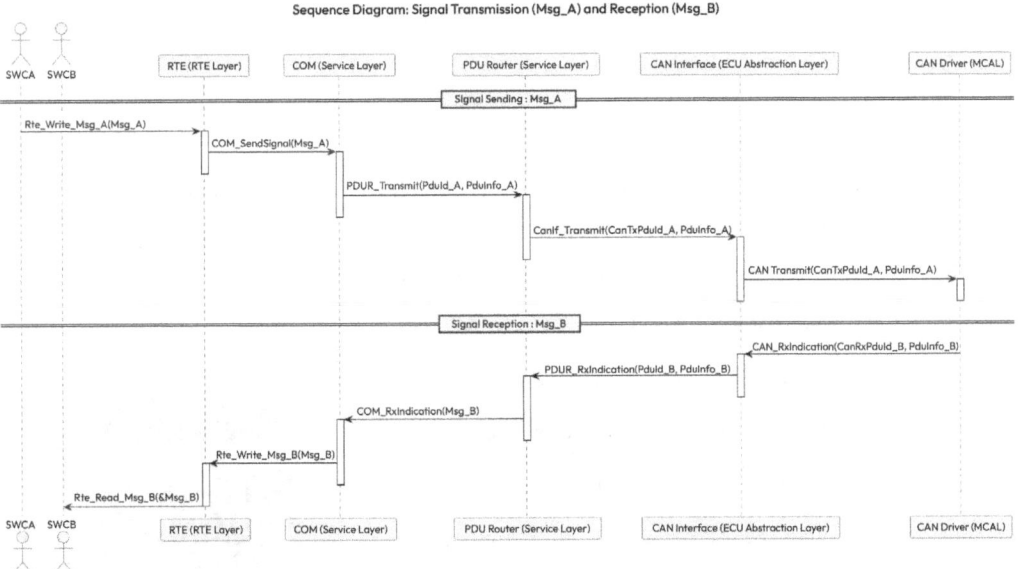

Figure 2.14 – Communication flow in AUTOSAR layers

This case study helps clarify the AUTOSAR architecture and showcases the benefits of its modular, standardized approach in a real-world application. By understanding the roles of the layers and the various interfaces, it becomes easier to grasp how the AUTOSAR framework facilitates efficient software development and integration within the automotive domain. This knowledge can be valuable when developing, integrating, or maintaining automotive software.

Summary

In this chapter, we explored the importance and relevance of various layers within the AUTOSAR architecture, emphasizing their role in automotive software system development. The insights gained in this chapter are essential, as they offer guidance on designing and implementing modular, reusable, and compatible SWCs that conform to the AUTOSAR standard.

Comprehending each layer, such as the application layer, RTE, service layer, ECU abstraction layer, and MCAL, is crucial for developers and engineers to effectively create automotive systems that are compliant with the AUTOSAR standard. This understanding allows them to efficiently utilize the services offered by each layer and ensures smooth interaction between SWCs and the underlying hardware.

By grasping the functions of different AUTOSAR layers through the mentioned use case of a battery management ECU, developers can create SWCs that are not only modular and reusable but also compatible with a variety of hardware platforms. This fosters adaptability and versatility within the automotive software ecosystem.

In the upcoming chapter, we will discuss the methodology of AUTOSAR, including how software and configuration are modeled and shared.

Questions

1. What are the AUTOSAR layers?
2. For the case study discussed in this chapter, try and match the requirements to the layers in which they could be implemented.

    Hint: One requirement could be matched with multiple layers.

3. What is the benefit of having different layers in AUTOSAR?

Get This Book's PDF Version and Exclusive Extras

Scan the QR code (or go to `packtpub.com/unlock`). Search for this book by name, confirm the edition, and then follow the steps on the page.

Note: Keep your invoice handy. Purchases made directly from Packt don't require one.

3

AUTOSAR Methodology and Data Exchange Formats

Having acquired knowledge about AUTOSAR and its various layers, we are now poised to delve into the methodology that underlies this innovative approach and where it fits in the automotive software development process, from requirements engineering to validation and testing. This knowledge will empower you to harness the full potential of the AUTOSAR framework. You will not only be able to apply it more effectively in your automotive software projects but also appreciate its significance in today's automotive industry.

In this chapter, we will attempt to understand this in depth by covering the following main topics:

- Introducing the AUTOSAR methodology
- Understanding the AUTOSAR methodology
- Using templates for data exchange
- Conformance classes for AUTOSAR

Introducing the AUTOSAR methodology

In this section, we will discuss the AUTOSAR methodology, outlining the steps or activities necessary for developing an AUTOSAR system. A key feature of the AUTOSAR methodology that distinguishes it from others is the substantial independence of software component implementation from **Electronic Control Unit (ECU)** Configuration. This separation offers the advantage of increased flexibility and adaptability, enabling developers to reuse software components across various ECU configurations and platforms, ultimately reducing development time and cost.

AUTOSAR Methodology and Data Exchange Formats

> **What is ECU Configuration?**
>
> **ECU Configuration** in AUTOSAR involves setting up both the software and hardware resources of an ECU to meet specific vehicle functions. This process includes selecting which AUTOSAR SWCs will be deployed on the ECU, establishing the foundational BSW layers, which handle communication, memory, and diagnostics, defining standardized protocols and interfaces for communication between the ECU, other ECUs, and external systems, and managing hardware resources, such as CPU, memory, and peripherals, to ensure the software components operate efficiently. This structured approach ensures that each ECU is tailored to perform its designated functions effectively within the vehicle's overall system.

While AUTOSAR offers various methodologies for different tasks, this discussion will focus on the most critical one, which involves the generation of a configured executable for each ECU based on software and hardware descriptions. This workflow is divided into three main parts, as shown in the following figure:

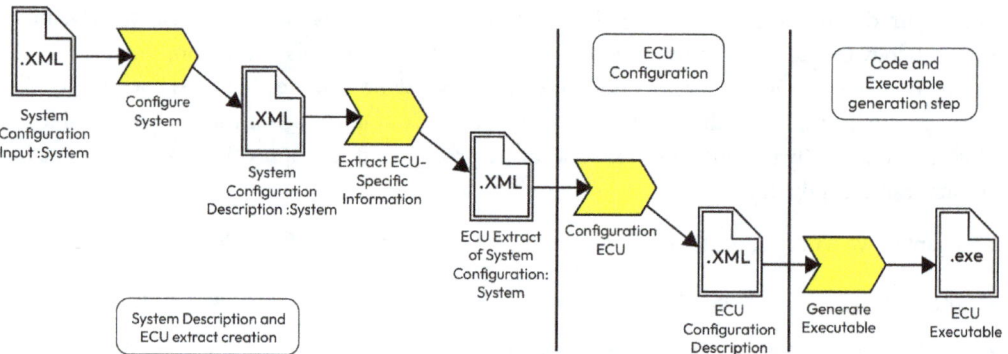

Figure 3.1 – AUTOSAR workflow

Let's look at these in detail:

- The first part of this workflow deals with template-based descriptions of hardware, software, and system constraints, resulting in a comprehensive system description that includes ECU software mapping.

- The second part details the configuration of the basic software and **runtime environment** (**RTE**) for a single ECU.

- The third phase involves generating the code and creating an executable for the ECU software. This executable is the outcome of compiling and linking various software components and basic software objects.

> **Note**
> While the methodology outlines the essential steps required to develop AUTOSAR systems, it does not describe a comprehensive process. For example, the methodology does not specify the frequency at which activities or activity sequences can be repeated, nor does it define roles or responsibilities involved in these activities.

As we continue our exploration of AUTOSAR, we now turn our attention to system configuration. This focuses on the configuration of software components and their interactions within the system, as well as their mapping to the corresponding ECUs.

Understanding the AUTOSAR methodology

The AUTOSAR methodology is a structured approach for retrieving the executables of ECUs in automotive systems. It consists of three main steps:

1. **System Configuration**: This step involves setting up the entire system. This includes defining the overall architecture, interconnections between ECUs, and specifying the communication protocols.
2. **ECU Configuration**: This step focuses on individual ECUs. In this step, specific information related to each ECU is extracted and prepared for subsequent actions. This information is essential for customizing the configuration of each ECU according to its requirements and functionalities.
3. **Code Generation**: In this step, the actual executable code for each ECU is generated. This code is derived from the configuration settings and specifications obtained in the previous steps. It ensures that each ECU operates according to its designated functionality within the system.

Let's look at these steps in greater detail.

System Configuration

An **original equipment manufacturer** (**OEM**) plays a pivotal role in developing an AUTOSAR system. The OEM is responsible for defining the system architecture, specifying the requirements, and overseeing the integration of software components from different suppliers.

> **Note:**
> In AUTOSAR, a *System Extract* represents a specific portion of a vehicle system, emphasizing certain aspects or configurations of particular ECUs. It originates from the comprehensive *System Description* and includes selected information about software components, interfaces, communication channels, and the mapping of components to ECUs.

To create the System Extract in an AUTOSAR system, an OEM typically performs the following steps:

1. **Define system requirements**: The OEM starts by defining the high-level requirements and functionalities for the automotive system. This may include performance, safety, security, and communication objectives.

   For example, in developing an **advanced driver assistance system (ADAS)**, the OEM sets the requirements for functionalities, such as adaptive cruise control, lane departure warning, and emergency braking.

2. **Identify software components and interfaces**: The OEM identifies the necessary software components and their interfaces based on the requirements. This may involve reusing existing components or developing new ones.

   For the ADAS, the OEM identifies components, such as sensor fusion, vehicle control, user interface, and corresponding interfaces.

3. **Design system architecture**: The OEM designs the overall system architecture, specifying the interaction between software components and mapping them to ECUs.

   The OEM determines how the sensor fusion component communicates with the vehicle control component and maps these components to specific ECUs within the ADAS.

4. **Develop or procure software components**: The OEM develops the required software components in-house or procures them from suppliers.

   For the ADAS, the OEM may develop the sensor fusion component internally while procuring the vehicle control component from a specialized supplier.

5. **Create system extract**: The OEM combines the software components, hardware topology, and configuration parameters into the System Extract.

   The System Extract for the ADAS includes the software components, ECU mapping, and configuration parameters for communication, sensors, and actuators.

The following figure illustrates an example of a System Extract involving multiple ECUs, showcasing the mapping of SWCs between them and the communication channels utilized by each ECU:

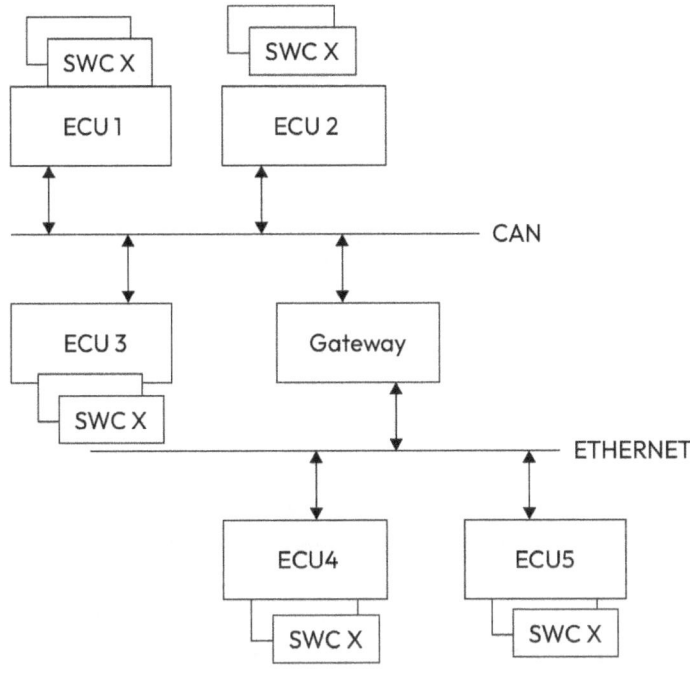

Figure 3.2 – Example of a system description

During the system configuration stage, the OEM assigns the software features to the ECUs within the system topology. While decisions about specific component implementations could be made at this point, these choices are typically not made at this high-level phase.

Additionally, a **communication matrix** is created by the OEM during this step, detailing the communication medium between ECUs, data, and associated properties, such as timing, checksum, and more. The outcome of this step is stored as the **System Configuration Description**. You can observe from *Figure 3.2* what would be the outcome of this step. Generally, a system consists of multiple ECUs and various communication channels. The communication between these ECUs is modeled, and if frames need to be gatewayed from one communication bus to another, an ECU can be the gateway. Numerous software tools are available to facilitate this process. With this understanding, let's proceed to the next step: configuring a specific ECU.

ECU Configuration

In most cases, the ECUs are manufactured by different suppliers for a specific vehicle. So, the individual ECU-specific information will be extracted from the *System Configuration Description* and given to the suppliers. This extract is called the **ECU extract**, as shown in the following figure:

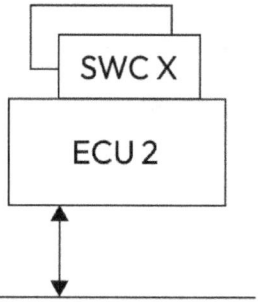

Figure 3.3 – Single ECU extract description

We use an AUTOSAR **authoring tool** to create a high-level design, which includes the Software Component Description, ECU Resource Description, and System Description files based on the templates discussed earlier (including the communication matrix, topology, and so on). Once we have these three components, the software development process using the AUTOSAR methodology can be largely broken into two pieces: **application/SWC software development** and the **ECU Configuration process**.

> **Note**
>
> AUTOSAR authoring tools are software applications designed to help create, modify, and visualize AUTOSAR system descriptions, software components, and their configurations. These tools play a vital role in the efficient development and management of AUTOSAR-based applications.

In the application SWC development process, the **SWC Description** file, which describes the software component elements, such as ports, interfaces, and triggering events that need to be developed, is taken as input; note that SWC elements shall be discussed in detail in later chapters.

We equip the ECU with all vital information necessary for implementation, including task scheduling, selection of appropriate BSW modules, configuration of the BSW, allocation of runnable entities to tasks, communication parameters, and various other configuration values that ensure optimal system performance. The output is an ECU Configuration Description file, which would contain the definition of micro-controller ports, pin assignment, clock config, and communication channel types, such as **controlled area network** (**CAN**), Ethernet, and FlexRay, in addition to specific service modules, such as **communication module** (**COM**) or specific peripherals such as **serial peripheral** interface (SPI), **universal asynchronous receiver-transmitter (UART)**, **inter-integrated circuit (I2C)**, or other peripherals.

The development of SWCs will be taking place in parallel with this step, which we will discuss in further detail in *Chapter 4*. The ECU Configuration Description file is used to generate the RTE and certain portions of the BSW. These components are then compiled along with the SWCs, ultimately creating an executable for a single ECU, as illustrated in the following figure:

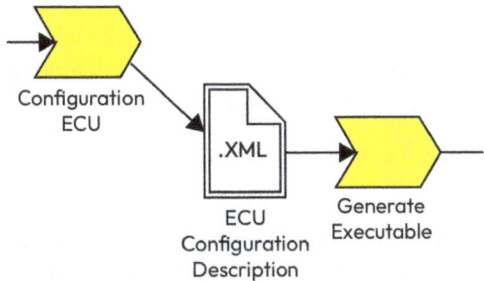

Figure 3.4 – Executable generation process

Moving forward, we will now introduce you to the code generation process in AUTOSAR. This essential step in the development cycle ensures efficient and accurate translation of configurations and templates into functional software.

Code Generation

The configuration tool generates code templates for the application software during this process. The interaction between these SWCs and the BSW layer determines the generation of the RTE code.

> Note
> In the context of AUTOSAR, configuration tools play a vital role in the software development process. These tools facilitate the generation of configuration files and code templates, allowing developers to customize software components and BSW modules. It is essential to be aware that various vendors offer their own configuration tools, each with distinct features and capabilities. Therefore, it is important for developers to choose the appropriate tool that best meets their specific requirements and aligns with their AUTOSAR-based development process.

For each BSW module, configuration files are created that encompass all user-defined configurations made within the AUTOSAR configuration tool. It is also important to note that these configurations include subsets of types. These are known as **configuration classes**, which refer to the stages in the development process at which specific parameters can be set or modified. These classes determine when and how configuration parameters are defined and when they become fixed during the software development process. Let's look at the differences between the three main configuration classes in AUTOSAR:

- **Pre-compile time**: Parameters of this configuration class are set before the compilation process begins. They are usually hardcoded or defined as macros or constants in the source code. These parameters cannot be changed after the compilation.

  An example use case for pre-compile time parameters is enabling a macro for error tracing during development.

- **Link time**: Parameters of this configuration class are used after the compilation but during the linking process. These parameters are usually specified as object code files and become fixed after the linking process, in contrast to pre-compile time configurations, which are established during the compiler preprocessor (e.g., the `#define` constants).

  An example use case for link time parameters is the usage of configuration structures that contain configurations of CAN communication, such as the channel identifier.

- **Post-build time**: Parameters of this configuration class can be changed after the entire build process (compilation and linking) is completed. There are two alternatives for post-build time parameters:

  - **Loadable**: These parameters are stored in a non-executable binary file that can be downloaded to an ECU. The parameter values can be changed during the boot process of the ECU.

  - **Selectable**: These parameters are stored in multiple alternative configuration sets linked to the executable. The configuration settings can be selected during the boot process of the ECU and can be downloaded to an ECU during the boot.

We have explored the three main configuration classes in AUTOSAR: pre-compile time, link time, and post-build time. Each class plays a role in determining when and how configuration parameters are defined and fixed during the software development process. By understanding these classes and their implications, developers can effectively manage and optimize the configuration of their AUTOSAR-based systems.

In the next section, we will be introduced to the templates of AUTOSAR. We will also understand how the software and system are modeled and shared between different stakeholders.

Using templates for data exchange

AUTOSAR templates serve as formalized representations of information structures utilized in the AUTOSAR methodology and are used to define software components, modules, or configuration parameters. Templates ensure consistency, modularity, and efficiency across different ECUs and vehicle platforms by providing predefined frameworks that developers can instantiate and customize based on specific application needs; they ensure uniformity, standardization, and smooth data exchange throughout different phases of the development process. These templates are specified using the **unified modeling language** (UML) and rely on the AUTOSAR **UML Profile**. The profile introduces a series of stereotypes that convey AUTOSAR concepts, such as the hardware elements of ECUs, software components, ports, data types, or interface descriptions. These stereotypes are systematically arranged into packages based on their application within diverse templates.

All templates rely on the AUTOSAR UML Profile. The template packages within the UML Profile are interrelated, as illustrated in the following figure. A connection between two packages indicates that one package reuses concepts defined in the other:

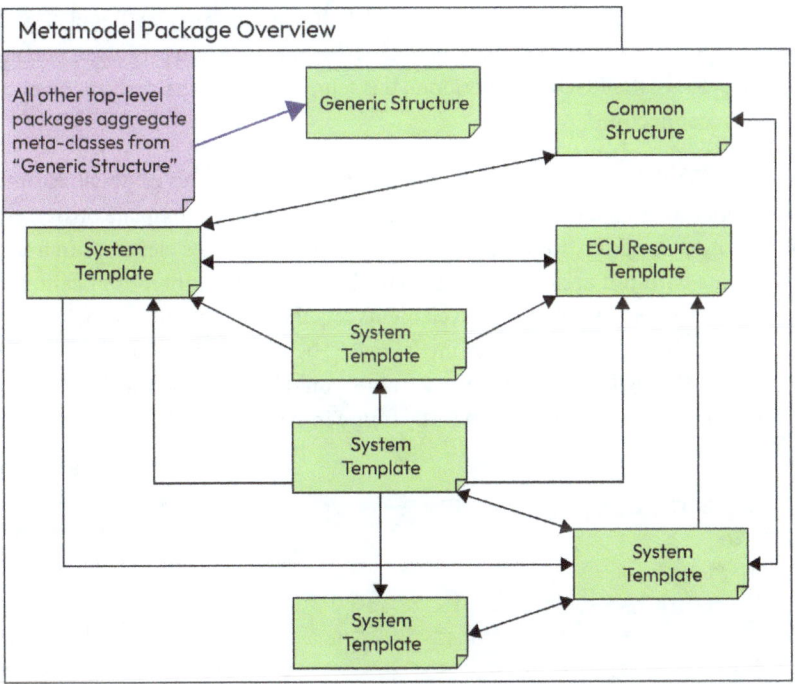

Figure 3.5 – AUTOSAR UML Profile

Let's look at the components shown in the preceding figure in greater detail:

- The **Software Component Template** describes the software aspect of the system. The template includes a general description of component types with port prototypes and port interfaces. In addition, it distinguishes between composite components, which leads to a hierarchical structure of the software components.

- The **ECU Resource Template** outlines the hardware resources of an ECU, such as those found in an **Engine Control Module** (ECM), including physical communication media and peripherals, such as ADCs and **General Purpose Input/Output** (GPIOs). The information provided is crucial for configuring the ECU abstraction layer and the microcontroller to tailor them explicitly for that ECU.

- The **Basic Software Module Description Template** addresses the elements of basic software. It encompasses all data related to basic software modules and clusters. There's a connection between this template package and the Software Component Template, as both detail implementation features and resource usage.

- The **System Template** serves as a link between the Software Component Template and the ECU Resource Template. It establishes a connection between the system's software perspective and the physical layout of the ECUs. Furthermore, this template allows the description of constraints regarding this connection.

Every template in AUTOSAR adheres to the **Generic Structure**, which establishes fundamental properties. For instance, one such property is that each element within a template possesses a distinct identifier or hierarchical organization; the Generic Structure organizes elements in a hierarchical manner, and elements within it are organized into packages based on their application in various templates. These packages help to group related elements together and maintain a logical organization within the AUTOSAR system. For example, in the following two snippets, we show how a CAN frame is modeled in **AUTOSAR XML** (**ARXML**); every frame would be described within the `CanFrame` parent node. Here is an example of defining a CAN frame in ARXML:

```
<AR-PACKAGE>
    <SHORT-NAME>CanFrame</SHORT-NAME>
    <ELEMENTS>
        <CAN-FRAME>
            <SHORT-NAME>CanFrameExample</SHORT-NAME>
            <FRAME-LENGTH>8</FRAME-LENGTH>
            <PDU-TO-FRAME-MAPPINGS>
                <PDU-TO-FRAME-MAPPING>
                    <SHORT-NAME>Cluster_0_Status_0</SHORT-NAME>
                    <PACKING-BYTE-ORDER>MOST-SIGNIFICANT-BYTE-LAST</PACKING-BYTE-ORDER>
                    <PDU-REF DEST="I-SIGNAL-I-PDU">/PDU/CanFrameExample</PDU-REF>
```

```
                    <START-POSITION>0</START-POSITION>
                </PDU-TO-FRAME-MAPPING>
            </PDU-TO-FRAME-MAPPINGS>
        </CAN-FRAME>
    </ELEMENTS>
</AR-PACKAGE>
```

The ARXML snippet defines a CAN frame named `CanFrameExample`, which has a length of 8 bytes. It includes a mapping of the `CanFrameExample` **protocol data unit** (**PDU**) to the CAN frame, starting at position 0, with the most significant byte placed last. The mapping is identified as `Cluster_0_Status_0`.

CAN frame triggering is one of the ways to describe how messages are sent and received between ECUs using the CAN bus:

```
<CAN-FRAME-TRIGGERING>
    <SHORT-NAME>FT_CanFrameExample</SHORT-NAME>
    <FRAME-PORT-REFS>
        <FRAME-PORT-REF DEST="FRAME-PORT">/ECU/MyECU/CAN/FP_
CanFrameExample_Rx</FRAME-PORT-REF>
    </FRAME-PORT-REFS>
    <FRAME-REF DEST="CAN-FRAME">/CanFrame/CanFrameExample_OCAN</FRAME-REF>
    <PDU-TRIGGERINGS>
        <PDU-TRIGGERING-REF-CONDITIONAL>
            <PDU-TRIGGERING-REF DEST="PDU-TRIGGERING">/Cluster/CAN/
CHNL/FT_CanFrameExample</PDU-TRIGGERING-REF>
        </PDU-TRIGGERING-REF-CONDITIONAL>
    </PDU-TRIGGERINGS>
    <CAN-ADDRESSING-MODE>STANDARD</CAN-ADDRESSING-MODE>
    <CAN-FRAME-RX-BEHAVIOR>ANY</CAN-FRAME-RX-BEHAVIOR>
    <CAN-FRAME-TX-BEHAVIOR>CAN-200</CAN-FRAME-TX-BEHAVIOR>
    <IDENTIFIER>1890</IDENTIFIER>
</CAN-FRAME-TRIGGERING>
```

This ARXML snippet outlines the triggering configuration for the `PT_CanFrameExample` CAN frame, detailing how it is linked to both the receive (`CanFrameExample_Rx`) and transmit (`CanFrameExample_oCAN`) ports of an ECU, and referencing its associated PDU (`PT_CanFrameExample`). It sets the addressing mode to `STANDARD`, defines specific receive and transmit behaviors, and assigns the frame a unique identifier (`1809`), which dictates how this CAN frame interacts within the CAN network during message communication.

In the next section, we shall explore the different types of conformance classes in AUTOSAR and how this ensures compatibility and interoperability within the AUTOSAR ecosystem.

Conformance classes for AUTOSAR

The **conformance classes** define a minimum set of BSW modules that an ECU must support to be considered compliant with a specific conformance class. By adhering to a conformance class, ECU manufacturers can ensure their products are compatible with other AUTOSAR-compliant systems and components. Let's list out the types of conformance classes:

- **Conformance Class 1 (CC1)**: This class represents the most basic level of AUTOSAR support, primarily focusing on essential features, such as communication and basic diagnostics. ECUs conforming to CC1 typically have limited resources and are used in simple applications.

- **Conformance Class 2 (CC2)**: ECUs conforming to CC2 provide more advanced features than CC1, such as support for complex device drivers, memory management, and advanced diagnostics. CC2 is suitable for ECUs with more resources and complex use cases.

- **Conformance Class 3 (CC3)**: This class is designed for high-performance ECUs with advanced capabilities, such as support for multicore processors, advanced scheduling, and complex communication protocols. CC3 is typically used in high-end, safety-critical applications.

- **Conformance Class 4 (CC4)**: CC4 represents the most advanced level of AUTOSAR support, incorporating features such as adaptive platform support, advanced cybersecurity, and support for new communication protocols. This class is intended for highly sophisticated ECUs that require cutting-edge technology and features.

In summary, we have explored the various conformance classes of AUTOSAR, which define the minimum set of BSW modules required for an ECU to be considered compliant. From the basic features in CC1 to the highly advanced capabilities of CC4, these classifications ensure compatibility among different AUTOSAR-compliant systems and components. With this understanding of the conformance classes, you are now better equipped to select the appropriate class for your specific ECU development needs and ensure seamless integration with other components in the automotive ecosystem.

Summary

In this chapter, we explored the AUTOSAR methodology, outlining a series of activities necessary for developing an AUTOSAR system, focusing on the independence of software component implementation from ECU Configuration. We also discussed the data exchange formats, specifically AUTOSAR templates, which are formalized representations of information structures used throughout development. Additionally, we delved into the process of system design and modeling, code generation and configuration, and the various conformance classes, which define a minimum set of BSW modules that an ECU must support to be considered compliant with a specific conformance class.

By understanding these core concepts, you can utilize the AUTOSAR methodology for system design, ensure smooth data exchange using AUTOSAR templates, and adhere to the appropriate conformance classes for compatibility with other AUTOSAR-compliant systems and components.

In the following chapter, we will begin our exploration of the software components and RTE as an initial step toward understanding more of the AUTOSAR building blocks.

Questions

1. Describe the steps involved in designing an automotive system with AUTOSAR methodology.
2. What are conformance classes for AUTOSAR?
3. Discuss the different configuration classes.

Get This Book's PDF Version and Exclusive Extras

Scan the QR code (or go to `packtpub.com/unlock`). Search for this book by name, confirm the edition, and then follow the steps on the page.

Note: Keep your invoice handy. Purchases made directly from Packt don't require one.

Part 2: Investigating the Building Blocks of AUTOSAR

This part delves into the core components of AUTOSAR, focusing on **software components** (**SWCs**) and the **runtime environment** (**RTE**). It explores the intricacies of events, interfaces, and the communication stack, including critical modules such as COM and PDUR. You will learn about the AUTOSAR operating system and its scheduling and task management features. By understanding the building blocks through several of the most important components, this section equips developers with the knowledge needed to design and implement robust, modular, and efficient automotive software systems.

This part has the following chapters:

- *Chapter 4, Working with Software Components and RTE*
- *Chapter 5, Designing and Implementing Events and Interfaces*
- *Chapter 6, Getting Started with the AUTOSAR Operating System*
- *Chapter 7, Exploring the Communication Stack*

4
Working with Software Components and RTE

This chapter aims to elucidate how **software components** (**SWCs**) serve as encapsulating units for various functions within the system and communicate with other SWCs or BSW modules with well-defined interfaces. We will try to explore the diversity of SWCs, shedding light on the different types of SWCs. By understanding the spectrum of these components, we aim to provide readers with a holistic understanding of their function and importance in the AUTOSAR context.

One of the key highlights of this chapter is the exploration of the AUTOSAR **runtime environment** (**RTE**). This critical aspect of the AUTOSAR framework serves as the communication hub for SWCs, ensuring data integrity and consistency during exchanges between the SWCs. We will guide you through its mechanics, giving you an understanding of how it facilitates effective communication within the system.

This chapter focuses on understanding SWCs and how they are being designed. It will cover the following topics:

- Understanding SWCs
- Exploring AUTOSAR datatypes
- Modeling an SWC
- Communication between the SWCs
- Understanding the significance of the RTE

First, let's explore what a software component really means.

Understanding SWCs

As discussed in *Chapter 2*, in the AUTOSAR standard, an SWC refers to an encapsulated piece of software with a specific function within an embedded automotive system.

An AUTOSAR software component is designed to be independent, reusable, and transferable between different **electronic control units (ECUs)** in a vehicle.

These SWCs are made up of a set of runnable entities (i.e., code functions) that are activated or scheduled by certain events and interact with the system through well-defined interfaces.

> **Important note**
> Imagine a car is like a big puzzle, and every part of the car needs to work together smoothly. In AUTOSAR, SWCs are like individual pieces of that puzzle. They are small, self-contained parts of the car's computer systems, each with a specific job to do. Some of these components are grouped together within a single ECU, while others are distributed across different ECUs throughout the vehicle.

The following figure illustrates the interaction and data exchange between SWCs via the RTE. Notice how the SWCs are positioned above the RTE, and the RTE is responsible for managing the exchange of data and control between them:

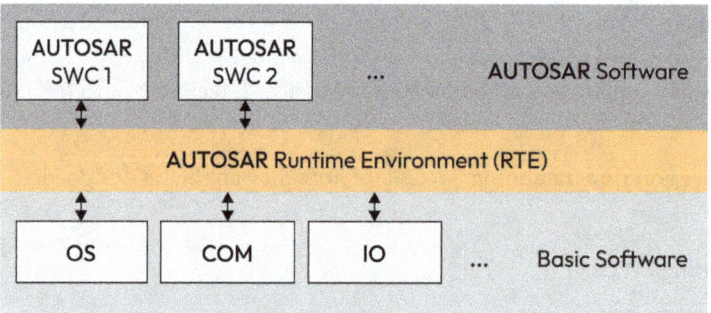

Figure 4.1 – SWCs in AUTOSAR

Some characteristics of AUTOSAR SWCs are as follows:

- Their capability to link with an AUTOSAR RTE is crucial. This link enables them to communicate with other SWCs and with other SWCs within the BSW layer of the AUTOSAR architecture. This results in multiple benefits like the reusability of software, improved mobility of components between ECUs, and more resourceful and economical utilization of resources.

- AUTOSAR SWCs are also scalable. They can form AUTOSAR compositions, a unique type of software component that allows grouping SWCs with similar functionalities. These compositions function as a system abstraction layer, handling complexity and aiding scalability in designing a software application's logical representation.
- AUTOSAR SWCs offer a modular, expandable, and efficient approach to automotive software development. They play a crucial role in the smooth operation of complex automotive systems, and their adaptability and reusability make them a valuable resource in the ever-evolving world of automotive software architecture.

Since the components can communicate seamlessly, the same software modules can be reused across different projects and platforms, reducing development time and effort. To understand SWCs better, let's look at an example.

Example – Throttle controller

Consider a *throttle control* component in an automotive engine management system. This component might be responsible for controlling the position of the throttle valve based on inputs from the driver's acceleration pedal, engine control unit, or cruise control system. The throttle control SWC might include several functions, such as the following:

- `calculateThrottlePosition()`: This function takes the input from the acceleration pedal sensor and calculates the desired throttle position.
- `actuateThrottle()`: This function takes the desired throttle position and controls the physical throttle valve accordingly.

Here's a simplified example of how these might look in C code:

```
void calculateThrottlePosition(AccelerationPedalPosition position) {
    // Algorithm to calculate throttle position based on pedal position
    ...
}
void actuateThrottle(ThrottlePosition position) {
    // Control signal sent to throttle actuator based on calculated position
    ...
}
```

In this example, the inputs and outputs would be passed through the defined interfaces of the SWC, allowing it to interact with other components such as a **sensor SWC** (for acceleration pedal sensor input) and an **actuator SWC** (for controlling the throttle actuator). *Figure 4.2* shall show a simplified example of such interaction.

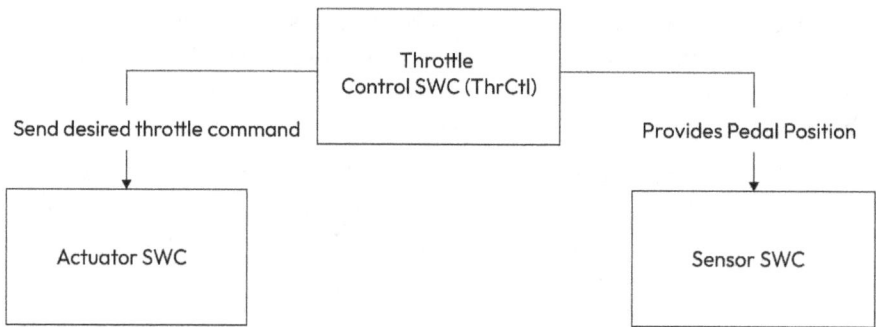

Figure 4.2 – Throttle Control SWC interaction

The AUTOSAR authoring tool would be used to create the descriptions of these components, runnable entities, and their interfaces, typically in a standardized XML-based format, **AUTOSAR XML (ARXML)**, which would then be used to generate the necessary glue code or wrapper layer for integrating these components into the overall system (ECU). This will be explored later on, when we explore the building blocks of the SWC.

We now have an understanding of what a SWC might do. A question might arise as to how it would integrate into AUTOSAR software. Additionally, we might wonder how these SWCs are organized, how they communicate, and how they are represented in models. First, let's delve deeper to understand the various types of SWCs, then figure out how to model and realize them into AUTOSAR.

Types of AUTOSAR SWCs

In this section, we will delve into an array of fundamental SWCs that form the bedrock of the AUTOSAR framework, explaining their unique roles and interactions within the intricate architecture:

- **Application SWC**: This represents an indivisible software component that performs a part or whole of an application. Confined to only one ECU, this component forms the fundamental building block of an AUTOSAR application.

- **Parameter SWC**: This is primarily used to access calibration parameters from the ECU and to supply calibration parameters to other SWCs. This component does not include internal behavior and provides parameters to SWC through a specific type of interface.

- **Service SWC**: Standard interfaces for direct communication with other BSW modules are offered, for example, a service from the COMM module to manage communication channels' states.

- **Complex device driver**: This generalizes the ECU abstraction component and can communicate with BSW modules. If there's a complex application that AUTOSAR's BSW architecture cannot implement, then the complex device driver component can be used to implement this non-AUTOSAR-defined functionality within the AUTOSAR ECU, more extending functionality.

- **NVBlock SWC**: This manages non-volatile memory within an ECU. It ensures important data, such as calibration parameters, persists across power cycles, enhancing system reliability by offering an interface for reading and writing data that survives restarts.

Moving forward, our next focus will be the procedure for designing a software component within AUTOSAR. We'll familiarize ourselves with the SWC elements next.

Elements of SWC

AUTOSAR SWCs are highly structured entities that comprise several essential elements, as shown in *Figure 4.3*:

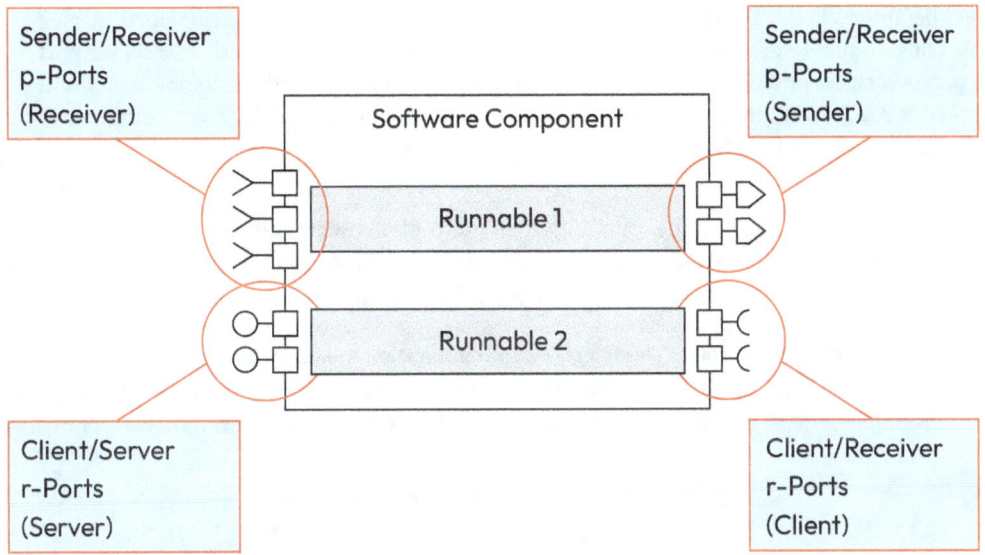

Figure 4.3 – Internal structure of a SWC

Let's look at these elements in detail:

- **Ports**: In AUTOSAR, a port acts as a well-defined *interaction point* between a SWC and the external world. It facilitates data exchange following a standardized approach. There are two types of ports in AUTOSAR:

    - **Provided ports (PPorts)**: This acts as an *output channel* for an SWC. Data is generated by the component and transmitted through the PPort to other SWCs or the **basic software (BSW)** layer, the PPort is like a service desk, offering specific services that other components or users. For example, a *Temperature Sensor* SWC may have a provided port that offers the service of reporting the current engine temperature.

- **Required ports (RPorts)**: This serves as an *input channel* for an SWC. The component expects to receive data through the RPort from other components or the BSW. It specifies the operations or functions that the component needs but doesn't implement by itself. In the context of our service provider analogy, the required Port is like a customer who requests specific services. For example, an *engine control* SWC may have a required port that needs the engine temperature. It will connect to the provided port of the *Temperature Sensor* SWC to get this information.

> **Note**
> Ports in AUTOSAR can be compared to electrical outlets and plugs in a house. Electrical outlets, like server ports in AUTOSAR, provide services or data to other components. Plugs, like client ports in AUTOSAR, request and receive services or data from server ports. A plug must fit an outlet to draw power, just as a client port must connect correctly to a server port to access services or data. Electrical outlets and plugs are standardized for compatibility, and similarly, AUTOSAR ports are standardized to ensure interoperability between components from different manufacturers.

- **Interfaces**: These establish the communication protocols between SWCs, enabling their smooth interaction. AUTOSAR provides several communication interfaces. However, in the current scope, we will focus on discussing the two most used categories of communication interfaces:
    - **Sender-receiver interface:** This refers to a communication model where one SWC, acting as the sender, transmits data to one or more SWCs acting as receivers. This form of communication is unidirectional, meaning data only flows from the sender to the receiver(s). It is represented as follows:

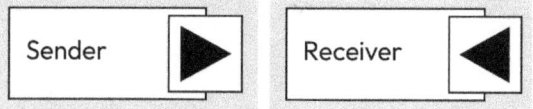

Figure 4.4 – Sender-receiver interface symbols

Consider an example as shown in *Figure 4.5*, where we have three SWCs in a vehicle's control system:

- **Speed sensor SWC (sender)**: This is responsible for detecting and recording the vehicle's current speed. The measured speed data is crucial information needed by other components in the system for different purposes. The Speed Sensor SWC (sender) sends this speed data to the Speed Display SWC and Cruise Control SWC (receivers) through the sender-receiver interface.

- **Speed display SWC (receiver)**: The Speed Display SWC uses this data to update the speedometer readout on the vehicle's dashboard, allowing the driver to see the vehicle's current speed.

- **Cruise control SWC (receiver)**: Meanwhile, the Cruise Control SWC uses the data to maintain or adjust the vehicle's speed per the driver's setting. For instance, if the vehicle's speed drops below the set speed, the Cruise Control SWC may instruct the engine control system to increase acceleration.

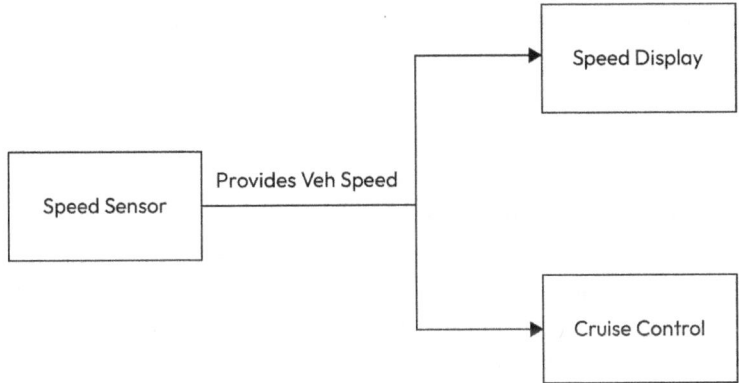

Figure 4.5 – Sender-receiver example

Having explored the sender-receiver interface, let's explore the client-server communication model in AUTOSAR:

- **Client-server interface**: It facilitates communication between SWCs based on a **request-response model**. Here, one SWC (the client) requests a specific service from another SWC (the server), and the server then responds by executing the service and sending a response back to the client. This is a bi-directional interface since communication occurs in both directions – from the client to the server and back. *Figure 4.6* shows how the client-server interface is represented in the AUTOSAR models:

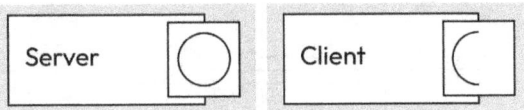

Figure 4.6 – Client-server interface symbols

For example, let's consider a vehicle's control system with two SWCs:

- Climate control SWC (client)
- Air conditioning system SWC (server)

The Climate Control SWC manages the vehicle's interior climate, including temperature, humidity, and airflow. To do this, it communicates with other systems in the vehicle, in this case, the Air Conditioning System SWC.

Here's how a client-server interface could work in this context:

i. **User input**: The driver adjusts the desired temperature on the climate control panel.

ii. **Climate analysis**: The Climate Control SWC receives user input and sensor data. Based on the difference between the desired and current temperature (and potentially humidity), the Climate Control SWC determines the necessary adjustments.

iii. **Client request**: The Climate Control SWC sends a request through its RPort, invoking a specific operation on the Air Conditioning System SWC (server). This operation could be named `SetTargetTemperature` and include arguments such as the desired temperature value.

iv. **Server operation**: The Air Conditioning System SWC receives the request and executes the `SetTargetTemperature` function with the provided argument.

v. **AC system control**: Based on the received target temperature, the Air Conditioning System SWC activates various actuators (e.g., compressor and fan) to adjust airflow and achieve the desired temperature.

In this example, the client-server interface within the AUTOSAR framework exemplifies how different SWCs can interact clearly and efficiently. This enables the system to respond precisely to external stimuli and execute the desired functionalities, ensuring a well-coordinated performance within the vehicle.

Other types of interfaces that are also used in AUTOSAR include the one shown in the following figure:

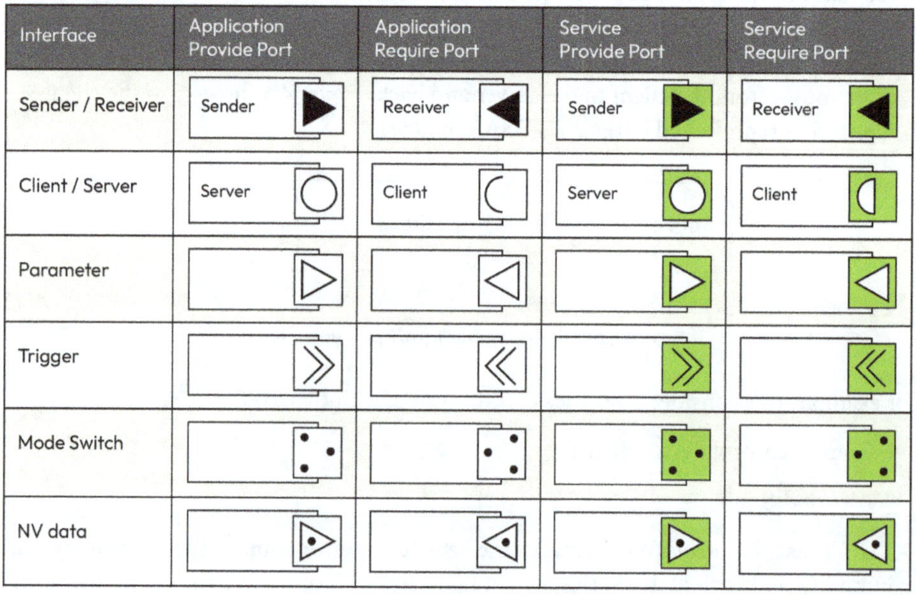

Interface	Application Provide Port	Application Require Port	Service Provide Port	Service Require Port
Sender / Receiver	Sender ▶	Receiver ◀	Sender ▶	Receiver ◀
Client / Server	Server ◯	Client ⊃	Server ◯	Client ⊃
Parameter	▷	◁	▷	◁
Trigger	≫	≪	≫	≪
Mode Switch	⸪	⸪	⸪	⸪
NV data	▷	◁	▷	◁

Figure 4.7 – List of AUTOSAR interfaces

> **Note**
> Interfaces in AUTOSAR can be compared to ethernet or telephone wires, which carry specific data, while ports can be seen as plugs that connect to these wires, enabling the exchange of data and services between components.

- **Runnable entities**: These encapsulate distinct functionality that the software component needs to perform. Each runnable entity corresponds to a specific task or operation of the component, such as processing inputs, calculating outputs, managing internal data, or executing specific algorithms. We can consider that a runnable is the implementation of a code function.
- **Internal behavior**: The internal behavior of an AUTOSAR SWC serves as a blueprint for its internal operations, meticulously detailing the fundamental building blocks called runnables, data access (defining how runnables consume or use data through ports and interfaces), and triggering mechanisms (initiating runnables based on data changes, time-based events, or event-driven triggers), thus offering a comprehensive picture of how the SWC functions internally.

These elements work together to ensure that AUTOSAR SWCs are scalable, reusable, and maintainable entities.

You've now grasped the basic concepts of SWCs, as well as their various types and elements. Furthermore, you've learned that a SWC is essentially a collection of different functions (runnables) that activate in response to events. Next, let's explore the various kinds of data that can be employed within these SWCs.

Exploring AUTOSAR datatypes

The AUTOSAR standard outlines a method for handling AUTOSAR data types, wherein base data types are associated with implementation and application data types. This approach allows for the distinction between physical attributes related to the **application level**, such as real-world value ranges, data structures, and physical meanings, and attributes specific to the **implementation level**, such as minimum and maximum values for stored integers and specifications of primitive types (e.g., integer, Boolean, or real).

Understanding this distinction is essential when working with SWCs because it ensures that the data is not only functional but also technically sound. It's like making sure that the puzzle pieces fit together perfectly, both in terms of their intended use and their technical compatibility.

Let's give a brief idea of what each data type represents:

- **Application data types**: In AUTOSAR, these are defined in a platform-independent manner, meaning they are not tied to a specific programming language or hardware platform. They capture the physical attributes, semantics, and behavior of data in the application domain. This includes information such as the range of valid values, units of measurement, data structures (e.g., records or arrays), and any relevant semantic constraints.

The purpose of application data types is to provide a standardized and portable representation of data within AUTOSAR systems. By separating the application-level attributes from the implementation-level details, application data types enable SWCs to be developed independently of specific hardware or software platforms. This promotes modularity, reusability, and interoperability within the AUTOSAR ecosystem.

- **Implementation data types**: In AUTOSAR, these are closely tied to the underlying programming language and hardware architecture. They are used to generate the actual code that will be executed on the target platform, considering aspects such as memory allocation, bit ordering, and alignment requirements. Examples of an implementation data type include the following:

    - `uint32`: This represents an unsigned integer with a size of 32 bits. This type is commonly used to store and manipulate non-negative integer values ranging from 0 to 4,294,967,295.

    - `float32`: This represents a single-precision floating-point number with 32 bits. It is used to perform calculations involving decimal values with a certain degree of precision.

    These implementation data types are the building blocks for defining the SWCs' variables, structures, and arrays. They provide the necessary detail and specificity to ensure compatibility and efficient execution on the target hardware and software platform.

- **Base types**: In programming, a base data type, also known as a primitive data type, is a fundamental data type provided by a programming language. Base data types are not composed of other data types and represent basic values that can be manipulated directly by the computer's hardware.

    Base data types encompass the hardware-specific characteristics of a data type, such as its size and encoding. They serve as the foundation upon which implementation data types are constructed. It is possible for multiple implementation data types to reference a single base type.

In the following figure, we can see how data types are connected. For instance, within the application, there's a data type known as **adt_VehSpeed**, and it's interesting to note that this data type has a maximum value set at 212. This illustrates how application data types in AUTOSAR carry specific attributes and constraints that are crucial for defining how data is used and processed within the SWC:

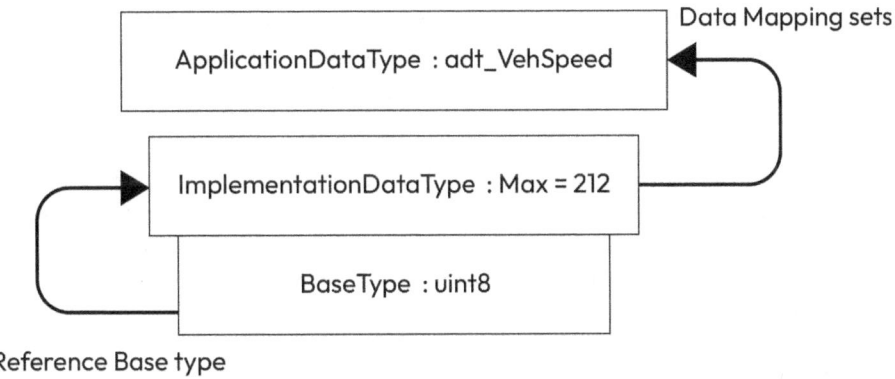

Figure 4.8 – Data types and their relation

Application data types are like labels that tell the software how to handle certain information. They define things such as the range of valid values, the structure of complex data, and what the data represents in the context of the application. In the case of `adt_VehSpeed`, it appears to pertain to vehicle speed, and knowing its maximum value is essential for ensuring the software component functions correctly when dealing with this type of data.

In the next section, we will explore how SWCs are modeled.

Modeling an SWC

Specialized software applications known as **authoring tools** streamline the creation and construction of SWCs and other aspects of the automotive software framework. These tools support engineers and developers in shaping, tailoring, and supervising various AUTOSAR elements, including SWCs and communication interfaces. Since tools such as Vector DaVinci Configurator and SystemDesk aren't available for free, let's explore the ARXML output instead. This will give us valuable insights into what to expect when using these tools.

What is an AUTOSAR authoring tool?

An AUTOSAR authoring tool is a specialized software application designed to interpret, manage, and create AUTOSAR descriptions, which are integral to the function and definition of SWCs. These descriptions highlight the functionalities, system requirements, resource demands, implementation specifics, and system constraints of SWCs, as well as details about ECU resources and configurations.

An SWC consists of two main parts: a comprehensive description that details its infrastructure setup, and an implementation that incorporates its functionality. The latter is usually written in C code. Before creating an SWC, its component type must be defined. This definition outlines its fixed characteristics, such as port names, communication properties, interface types and their specific attributes, and the SWC's behavior.

AUTOSAR elements interact with each other in a standardized XML file format, the ARXML format, which can vary slightly according to the AUTOSAR release version. Therefore, AUTOSAR authoring tools must be capable of interpreting, creating, or modifying ARXML descriptions based on the ARXML Schema version.

Let's consider this as an exercise to model a SWC. We will describe the process of designing an AUTOSAR software component named `TemperatureHandler`, which handles temperature data. In this example, it uses a sender-receiver interface to propagate temperature data. It also uses a client-server interface to query the temperature value (operation) from a server (provider). The definition of the software component in an ARXML file is as follows:

```
<AUTOSAR xmlns="http://autosar.org/schema/r4.0" xmlns:xsi="http://www.
w3.org/2001/XMLSchema-instance" xsi:schemaLocation="http://autosar.
org/schema/r4.0 AUTOSAR_4-3-0.xsd">
<AR-PACKAGES>
    <AR-PACKAGE>
    <SHORT-NAME>TemperatureControlPackage</SHORT-NAME>    <ELEMENTS>
............................
........................
<SW-COMPONENT-PROTOTYPE>
<SHORT-NAME>TemperatureHandler</SHORT-NAME>
    <PORTS>
        <P-PORT-PROTOTYPE SHORT-NAME="TempOutputPort">
            <PROVIDED-INTERFACE-TREF DEST="SENDER-RECEIVER-
INTERFACE">/TemperatureOutputInterface</PROVIDED-INTERFACE-TREF>
        </P-PORT-PROTOTYPE>
        <R-PORT-PROTOTYPE SHORT-NAME="TempInputPort">
            <REQUIRED-INTERFACE-TREF DEST="CLIENT-SERVER-INTERFACE">/
TemperatureSetInterface</REQUIRED-INTERFACE-TREF>
        </R-PORT-PROTOTYPE>
    </PORTS>
</SW-COMPONENT-PROTOTYPE>
<INTERNAL-BEHAVIORS>
<SWC-INTERNAL-BEHAVIOR>
    <SHORT-NAME>TemperatureControlSWC_Behavior</SHORT-NAME>
    <RUNNABLES>
    <RUNNABLE-ENTITY>
        <SHORT-NAME> TemperatureHandler_MainFunction
            </SHORT-NAME>
............................
........................
  </ELEMENTS>
    </AR-PACKAGE>
    </AR-PACKAGES>
</AUTOSAR>
```

In the preceding code snippet, the `TemperatureHandler` software component has been defined with two ports. The sender-receiver port is named `TempOutputPort`, and the client-server port is named `TempInputPort`.

The code for this software component might look like this:

```
#include "Rte_TemperatureHandler.h"
float currentTemperature = 0.0;
void TemperatureHandler_MainFunction(void)
{
    Rte_ResultType res = Rte_Call_TempInputPort_GetTemperature(&currentTemperature, &isupdated);
    if(isupdated == TRUE) {
        /* Propagate the new temperature */
        Rte_Write_TempOutputPort_Temperature(&currentTemperature);
    }
}
```

In this code, there is a main function (`TemperatureHandler_MainFunction`) that runs periodically. It checks whether a new temperature has been set. If so, it updates the temperature and propagates the new temperature data.

The following snippet is the generated header file that contains the API definitions:

```
#include "Rte_Type.h"

/* RTE API prototypes */
Std_ReturnType Rte_Call_TempInputPort_GetTemperature(float* temperature, boolean *isupdated);
Std_ReturnType Rte_Write_TempOutputPort_Temperature(const float* temperature);
```

In this RTE code, you can see the function prototypes for the RTE APIs that the software component uses. The `Rte_Call_TempInputPort_GetTemperature` function is used to check whether a new temperature value was updated. The `isupdated` flag indicates that a new temperature value has been updated and `Rte_Write_TempOutputPort_Temperature` is used to propagate the new temperature data to other users.

> **Note**
>
> Please note that the process would be more complex in a real-world application and involve additional steps. The code and ARXML snippets are very simplified for illustrative purposes and only show the port definition part. The code design is only to show how the modeling process would generate code.

Working with Software Components and RTE

At this point, we have a clearer idea of how a SWC design and implementation would work. Let's now examine how different SWCs would interact.

Communication between the SWCs

The **virtual functional bus** (**VFB**) implements communication between SWCs. It provides an abstraction layer that isolates the applications from the base infrastructure. During operation, the VFB is generally embodied by the RTE and is uniquely produced for each ECU in the AUTOSAR system. Communication within the VFB occurs through dedicated ports, requiring the mapping of communication interfaces from the application software to these ports. The VFB handles communication both within an individual ECU and between multiple ECUs, enabling a seamless exchange of data and information across the system:

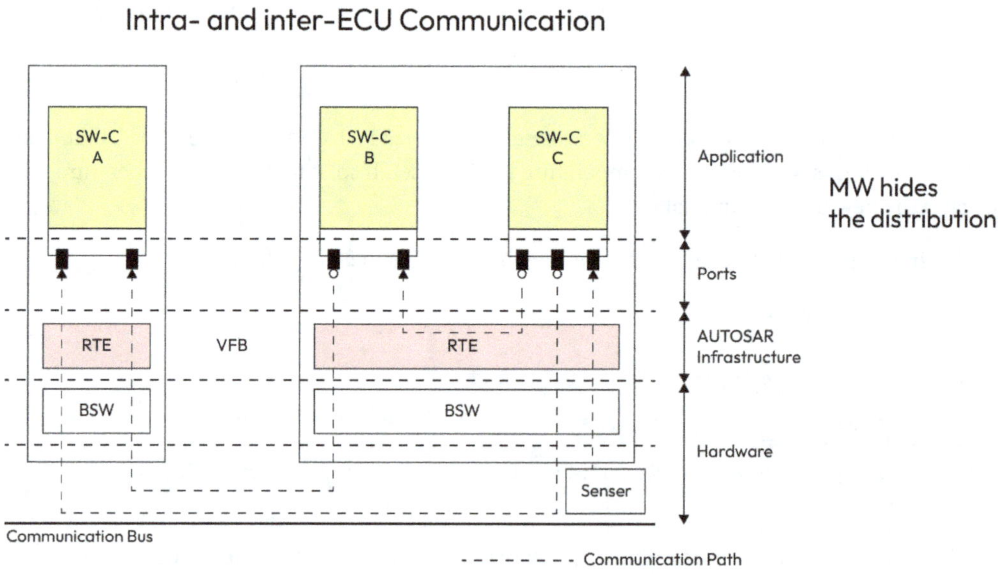

Figure 4.9 – Intra/inter-ECU communication

The communication between SWCs can occur in two ways, as depicted in the preceding diagram: inter-ECU and intra-ECU. Both types of communication are managed through the AUTOSAR RTE. Let's take a closer look at them:

- **Inter-ECU communication**, such as between *ECU I* and *ECU II*, occurs through and beneath the RTE, passing via the BSW module. The BSW module handles various functions, including memory interactions, diagnostics, and communication services (if necessary) to facilitate inter-ECU communication.

- **Intra-ECU communication** refers to communication within a single ECU, between software component B (SWC B) and software component C (SWC C). Intra-ECU communication is entirely handled through the RTE. The diagram shows that SWCs mapped to a single ECU utilize the intra-ECU communication method.

The application layer (Application SWCs) communicates with the lower layers and with each other via the RTE. Whenever an SWC needs a service, the RTE maps these requests to the appropriate service BSW service provider.

Now that you understand how SWC communication works, it's important to be introduced to a grouping for SWCs called **compositions**, which helps in managing complexity, logically organizing the system, and enhancing scalability.

Introducing compositions

SWCs and compositions are significant constructs in the AUTOSAR architecture, each serving a unique purpose. They offer different levels of abstraction and organizational methods to handle the complex ecosystem of automotive software.

We have explored SWCs earlier in this chapter; now let's understand compositions. These are a special type of SWC that doesn't contain any implementation of their own. Instead, they serve as containers to aggregate related SWCs, aiding in managing complexity and structuring larger systems.

For instance, you might have an *Engine System* composition that groups the *Engine Control*, *Fuel Sensor*, *Temperature Sensor*, and *Throttle Position* SWCs. The composition doesn't perform any functionality itself but aggregates these related components, improving the overall system's manageability. The creation of compositions offers distinct benefits:

- **Managing complexity**: As the number of SWCs in a system grows, managing them can become increasingly complex. Compositions provide a means to group related SWCs together, simplifying the architecture.
- **Logical organization**: Compositions help to logically organize the system by grouping related functionalities, enhancing the understandability of the system's structure and functions.
- **Scalability**: As a system expands or new features are added, new SWCs can be easily integrated by adding them to the appropriate composition.

In conclusion, the introduction of SWCs and compositions is significant in managing complexity and improving the development efficiency of automotive software systems. Now that we've covered SWCs and their compositions, it's time to explore how information moves between these entities via ports. This emphasizes why it's crucial to grasp the various **connector** types, which shall be explored next.

Connector types

Connectors serve as representations of data flow between PPorts and RPorts of SWCs or software compositions. They establish the necessary links for data exchange and coordination. In particular, *Assembly* and *Delegation* connectors are fundamental components within the AUTOSAR architecture that play a critical role in enabling communication and managing information flow between SWCs. Let's look at these in detail:

- **Assembly connector**: An Assembly connector establishes a direct link between a provided interface of one SWC and the required interface of another SWC. This connection enables seamless communication and information exchange between the connected components.

  For instance, let's consider an example involving a vehicle's cruise control system. This system comprises two SWCs: *Cruise Control Manager* and *Throttle Controller*. The *Cruise Control Manager* SWC offers a provided interface that outputs the desired speed, while the *Throttle Controller* SWC requires this desired speed information to control the throttle. To establish communication between these two components, an Assembly connector is employed. It connects the provided interface of the *Cruise Control Manager* SWC to the required interface of the *Throttle Controller* SWC, allowing the former to transmit the desired speed information directly to the latter. This connection ensures effective coordination and control within the cruise control system.

- **Delegation connector**: A Delegation connector is utilized within a composition to establish a connection between an inner SWC interface and an interface of the composition itself. It enables the composition to delegate the required or provided interface to one of its internal components.

  Expanding on the previous example, let's introduce a composition called *Vehicle Control System*, which encompasses the *Cruise Control Manager* and *Throttle Controller* SWCs. In this scenario, we may want the *Vehicle Control System* SWC to offer an interface to the external environment that provides the current speed information. To achieve this, we would use a Delegation connector to connect the *Vehicle Control System* interface responsible for outputting the current speed to the *Cruise Control Manager* SWC's interface, which is responsible for providing this information. The Delegation connector allows the *Vehicle Control System* SWC to delegate the task of providing the current speed interface to the *Cruise Control Manager* SWC, enabling access to this functionality through the composition interface.

Figure 4.10 illustrates the arrangement of SWCs and compositions, demonstrating how connections are established between them:

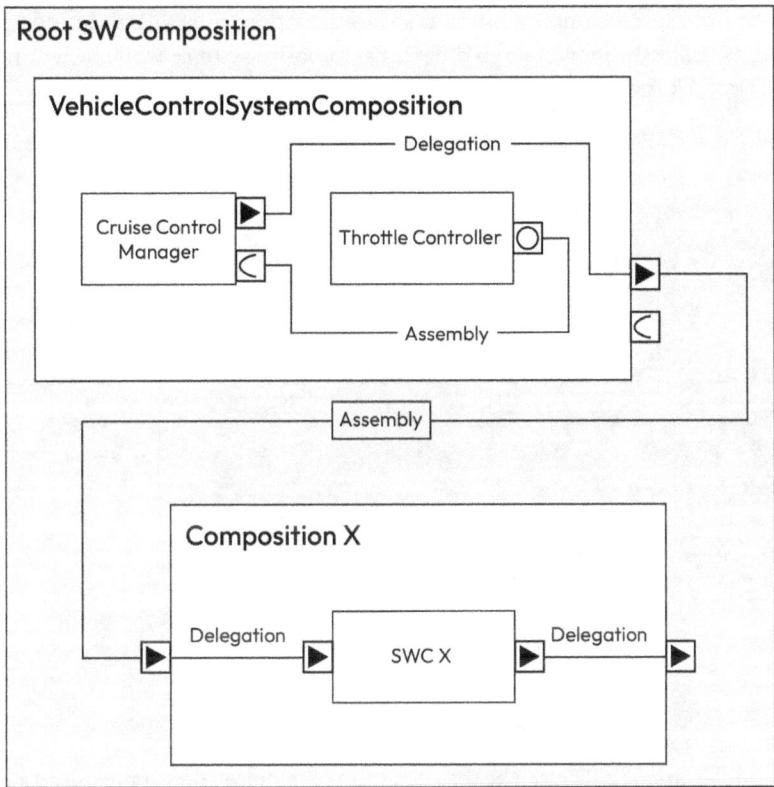

Figure 4.10 – Compositions and SWC

Some of the key differences between the Assembly connector and the Delegation connector are as follows:

- An Assembly connector connects a provider and receiver port. The Delegation connector connects the same port type, *provider to provider* or *receiver to receiver*.
- The Assembly connector facilitates the flow of data or signals directly between the connected SWCs. In contrast, the Delegation connector is meant to abstract the functionality of an inner SWC and expose it as part of the composition's interface. It allows the composition to mediate or provide specific functionalities, delegating the implementation details to the internal components.

Having observed the various types of communication between SWCs, it is now essential to delve deeper into the functionality and significance of the RTE layer.

Understanding the significance of the RTE

The collaborative efforts of the RTE and VFB enable efficient utilization of the underlying BSW layer, decoupling SWCs from specific implementations. This separation fosters flexibility and portability, allowing SWCs to seamlessly integrate into different automotive systems while benefiting from the standardized AUTOSAR framework.

Figure 4.11 shows how RTE acts as a middleware between the Application layer and the BSW layer.

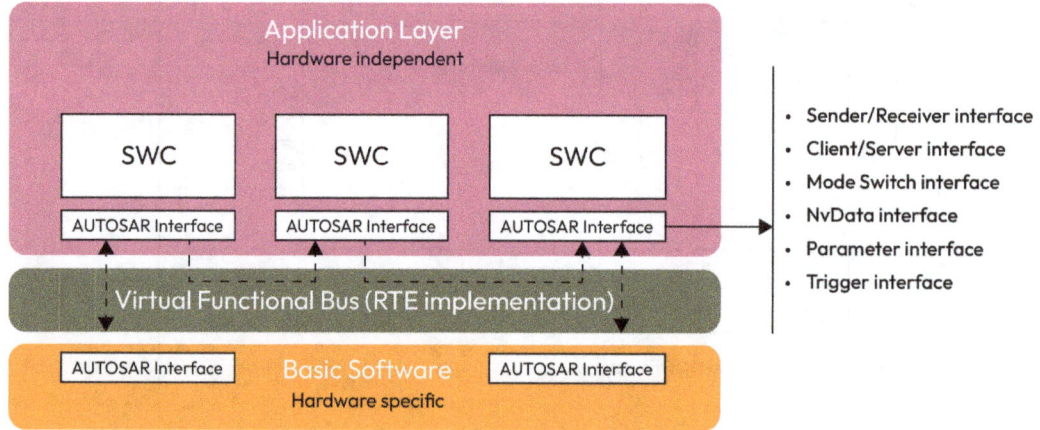

Figure 4.11 – RTE as a middle layer

The main features of the RTE layer are as follows:

- **RTE as middleware**: The RTE can be seen as a form of middleware that provides a standardized interface for SWCs. It abstracts the complexities of lower-level operations, such as data exchange mechanisms, task scheduling, and network management, presenting a simplified interface to the application layer. This allows application developers to focus on the core business logic while minimizing concerns about the underlying hardware or system-level tasks.

- **Communication management**: The RTE handles inter-ECU (between different ECUs) and intra-ECU (within the same ECU) communication. It efficiently manages the routing of information between sender and receiver components, irrespective of whether they reside within the same or different ECUs.

- **Data mapping and conversion**: The RTE is responsible for any necessary data mapping and conversion. As data is exchanged between components, it often needs to be transformed from one format to another. This may involve changing data types or encoding schemes.

- **Safety and security**: Due to its central role, RTE is critical in maintaining system integrity. It manages access rights, ensuring that only authorized components can communicate with each other. Furthermore, the RTE assists in identifying or mitigating specific faults such as timing errors and communication failures. It uses mechanisms such as error handling and monitoring to detect anomalies and either correct them or put the system into a safe state.
- **SWC–BSW interaction**: The RTE bridges the gap between the application layer (SWCs) and the BSW layer. SWCs interact with the BSW modules (such as OS and communication services) through standardized APIs provided by the RTE.
- **Event handling**: The RTE manages the execution of SWC functionalities. It can trigger specific parts of an SWC (runnables) based on pre-configured events, such as receiving data or a signal. This ensures tasks are executed efficiently.

Overall, the RTE is a critical component in the AUTOSAR architecture, performing essential operations that are key to both application functionality and system and software design efficiency. This sets the foundation for understanding how the RTE layer is instantiated in the system. In essence, the RTE serves as the binding layer that ensures seamless communication and interaction between various SWCs within the AUTOSAR framework.

Having grasped the fundamental operations of the RTE, the next question that arises is how this layer comes into existence.

Generation of RTE

The creation of the RTE within an AUTOSAR system is a structured procedure that converts the software architecture into a functional framework for the ECUs. Here's a general overview of how it operates, including an example and details on some of the critical output files.

The generation of the RTE utilizes various tools and input files to transform the AUTOSAR VFB specifications into operational code that functions on an ECU. This process ensures that all SWCs are capable of interacting with one another and with fundamental software modules, such as the **Operating System (OS)** and communication services, exclusively through interfaces defined by AUTOSAR.

For example, the input for RTE generation would be as follows:

- **SWC descriptions**: This includes information about the SWC's ports (data interfaces), interfaces (communication mechanisms like client-server), runnables (functional units), and events (triggers for execution).
- **ECU configuration**: This details the BSW modules available on the target ECU, such as the OS, communication stacks, and any of the BSW-provided services.
- **Toolchain**: This refers to the software tools used for RTE generation. Popular options include tools from AUTOSAR or third-party vendors such as Vector DaVinci Configurator.

> **RTE non-reusable nature**
>
> Due to its configuration-specific nature, the RTE is not reusable across different ECUs or projects. If the SWC deployment or communication patterns change, the RTE needs to be regenerated to reflect those changes.

Phases of RTE generation

The generation of the RTE in AUTOSAR is a two-phase process that transforms the high-level configuration of your SWCs into the specific code that manages communication and execution on the target ECU. These phases work together to ensure a tailored RTE for your system:

1. **RTE contract phase**: In this initial phase, the interactions between SWCs and the RTE are defined first. Limited information about the components' capabilities and communication needs is available. The main output of this phase is the application header file, which establishes a *contract* between the SWCs and the RTE. This contract specifies how the components will interact with the RTE and outlines the necessary interfaces and data types.

2. **RTE generation phase**: Following the contract phase, the RTE generation phase commences once all the relevant information about the SWCs, their deployment on ECUs, and their communication links is finalized. In this phase, the actual RTE code is generated. This code integrates all the specified interactions and ensures that the communication across SWCs, as well as between SWCs and BSW modules, adheres to the defined AUTOSAR standards.

The RTE isn't hand-coded, but rather automatically generated based on your project's specific configuration. Here's a look at the tools that handle this process.

Tools for RTE generation

A variety of specialized tools are used to support the RTE generation process. These tools are crucial for configuring, integrating, and generating the SWCs and RTE. They assist in establishing mappings, defining configurations, and ultimately producing the RTE code that will be implemented in the ECU. Here are the toolchain options:

- Vector DaVinci Configurator
- ETAS RTA-RTE
- EB tresos
- KPIT K-SAR

All these tools a powerful RTE generators that take ARXML files as input and produce the RTE code. It ensures that the generated code complies with AUTOSAR standards.

Choosing the right tool

The selection of the best tool depends on your specific needs and preferences. Here are some factors to consider:

- **Project requirements**: Some tools might cater to specific ECU architectures or offer advanced features for complex communication scenarios
- **Vendor support**: Consider the level of support and training offered by the tool vendor
- **Integration with existing workflow**: If you're already using other development tools, ensure compatibility and smooth integration with the chosen RTE generation tool
- **Cost and licensing**: Both AUTOSAR and third-party tools may have different licensing models and pricing structures

Having selected the right tools, let's now review how the RTE output will be generated. Note that all vendors will produce similar files, as the generated file structure and content are largely defined by the AUTOSAR standard. Although there may be slight variations in implementation, these files will have more similarities than differences.

Outputs of RTE generation

The output of the RTE generation process includes several key files that are critical for the software to function correctly within the ECU:

- Rte.c: This file contains the implementation code for the RTE, including the necessary function definitions, event handling, and data transformations.
- Rte.h: This file sets out fixed elements that don't need to be generated for every ECU, and are therefore not repeatedly generated by the RTE generator software. However, this file can be customized as per our application requirements if necessary. The Std_Types.h file is included in this file.
- Rte_<swc_name>.h: This file is a header generated, specific to each SWC. It defines the interface between the SWC and the RTE, facilitating communication and service access. This file houses the function prototypes of RTE APIs, data structures, and function prototypes of runnables used in the associated SWC. Rte_Type.h: This file is populated with RTE-specific type declarations that are based on the implementation data type configured during the SWC setup process. It also houses AUTOSAR data types that are beneficial for RTE APIs. The Rte.h file is included within this file.
- Rte_<swc_name>_Type.h: This file, often called the *Application types header file*, houses the constants related to the application, such as range values or enumeration values utilized within the SWC. The SWC name is always included within the file name as a spacer. The Rte_Type.h file is included in this file.

Additional files may be generated depending on the specific requirements and configurations of the project, such as RTE configuration files or additional header files for specific SWCs or modules.

Figure 4.12 shows the RTE generation steps with respect to SWC and RTE generator software:

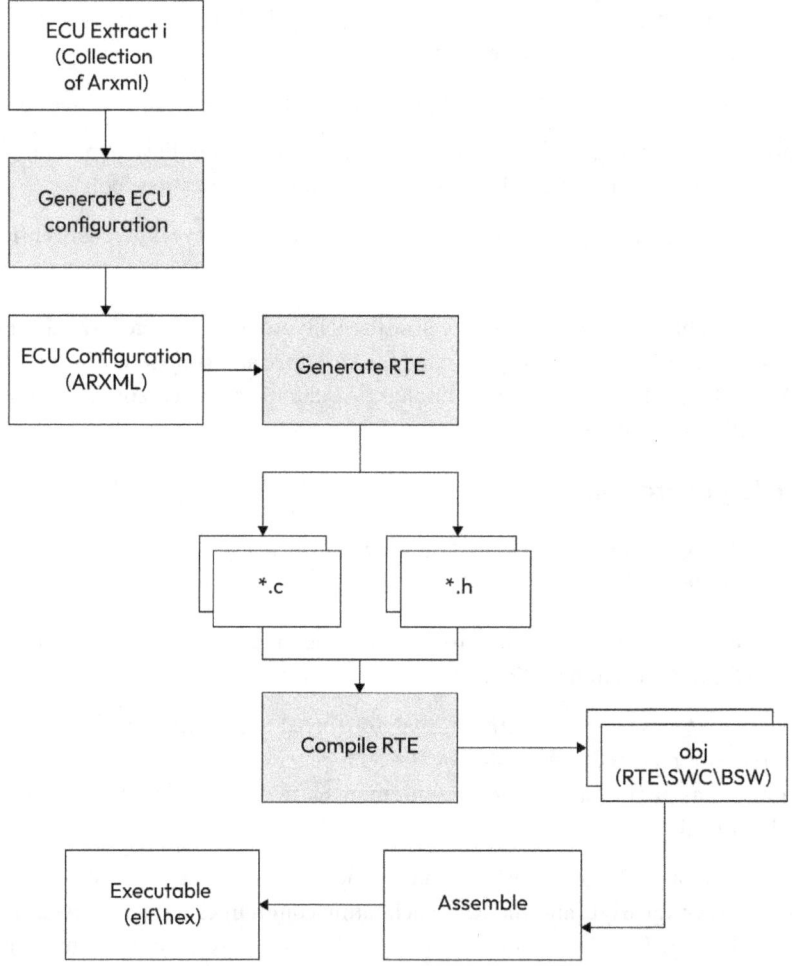

Figure 4.12 – Output RTE generation

Here is an overview of how this process goes:

1. **Extract ECU-specific information**:

    Process: ECU-specific information is extracted from the system configuration description.

    Output: This information is used to generate individual ECU configurations.

2. **Generate the ECU configuration:**

    Process: The ECU configuration values are generated based on the extracted information.

    Output: The ECU configuration values are stored in ARXML files.

3. **Generate RTE:**

    Process: The RTE is generated from the ECU configuration values.

    Output: This step produces RTE code files (`.c` and `.h`).

4. **Compile RTE:**

    Process: The generated RTE code is compiled.

    Output: RTE object files (`.obj`) are compiled.

5. **Compile SWCs and BSW:**

    Process: The SWCs and BSW are compiled.

    Output: SWC object files and compiled BSW object files (`.obj`) are compiled.

6. **Generate Executable:**

    Process: The compiled RTE, SWCs, and BSW object files are linked together to generate the final executable.

    Output: The ECU executable file (`.elf`/`.hex`/`.bin`, etc.) is the output.

We've now come to the conclusion of this chapter, where we attempted to demystify the intricate relationship between SWCs and RTE. It's important to note that we couldn't cover all the intricate details since this topic is quite extensive and goes beyond the scope of an introductory book. At this point, I hope you've gained an understanding of what SWCs are, how they integrate within an ECU, and their interplay within compositions, as well as the significance of ports and the methods by which these components connect.

Summary

This chapter dove into the world of AUTOSAR SWCs and their critical role in the AUTOSAR framework. We explored the structure, functionality, and key components of SWCs, gaining a solid foundation in the elements that drive automotive ECUs. SWCs were introduced as encapsulated, reusable units tailored to perform specific functions within an ECU. We also developed a thorough understanding of various SWC types, such as Application SWCs and complex device driver SWCs.

Our exploration revealed how SWCs use ports to communicate and interact, enabling the smooth exchange of services and data. We further examined how specific tasks and operations are structured within components through runnable entities and their triggering conditions.

Additionally, we understood the significant role of the RTE and its features, highlighting its essential connection with the VFB. With the introduction to fundamental SWC building blocks such as ports, interfaces, and internal behavior that this chapter has provided, developers are equipped to conceptualize complex systems.

In the next chapter, we will explore the design and implementation details of events and interfaces. We'll explore how ARXML models are structured and examine the internal characteristics of these interfaces, providing another layer of understanding of their role and functionality in the automotive software architecture.

Questions

1. What is a software component?
2. How is a SWC modeled in AUTOSAR?
3. What does internal behavior describe?
4. Discuss the function and features of RTE.
5. What is the difference between a sender-receiver interface and a client-server interface?
6. What does a port mean in AUTOSAR?
7. Describe the generation process for RTE and what inputs are needed.

Get This Book's PDF Version and Exclusive Extras

Scan the QR code (or go to `packtpub.com/unlock`). Search for this book by name, confirm the edition, and then follow the steps on the page.

Note: Keep your invoice handy. Purchases made directly from Packt don't require one.

5
Designing and Implementing Events and Interfaces

The interplay of **events** and **interfaces** constitutes the backbone of system communications and synchronization. In this chapter, we intend to dive deeper into how **software components** (**SWCs**) envelop diverse functions within the building blocks of automotive software, and employ events and interfaces to establish communication channels. Building on the foundation discussed in *Chapter 4*, we will explore the essential conduits of communication within SWCs: **ports**, **interfaces**, and **events**. We will unravel how they function and look at their various types, as well as their role in ensuring seamless data flow, providing a comprehensive insight into their roles and functionalities. Through this, you will gain a deeper understanding of the mechanisms that enable efficient and reliable communication in software systems.

We will be covering the following main topics:

- Introducing the communication model
- Understanding events and interfaces in AUTOSAR

Introducing the communication model

In the previous chapter, we discussed the concept of the **runtime environment** (**RTE**) a bit and highlighted the significance of SWCs as the foundational elements of any application within the AUTOSAR framework. We touched upon the idea that these components communicate and exchange information through specific channels known as ports and interfaces, which dictate the nature and format of the data being transferred.

In this section, we will try to understand the communication aspects within AUTOSAR. We will be discussing the communication model's key components, such as communication ports, interfaces, and the mechanisms through which SWCs interact with each other. This section will provide you with an understanding of how information flows within the AUTOSAR framework.

Explaining the information flow

Let's consider a simple embedded example. Imagine a very simple temperature monitoring component within an ECU. This system reads the temperature from a sensor, processes the data, and then displays the temperature on the car's dashboard by sending a CAN message to the infotainment ECU.

To translate this continuously varying analog signal into a format that software can understand, an embedded device would utilize an **analog-to-digital converter** (**ADC**). The ADC periodically converts the signal into digital values and provides it to the users, such as another software module.

Figure 5.1 shows briefly how such a system would look from an abstract view:

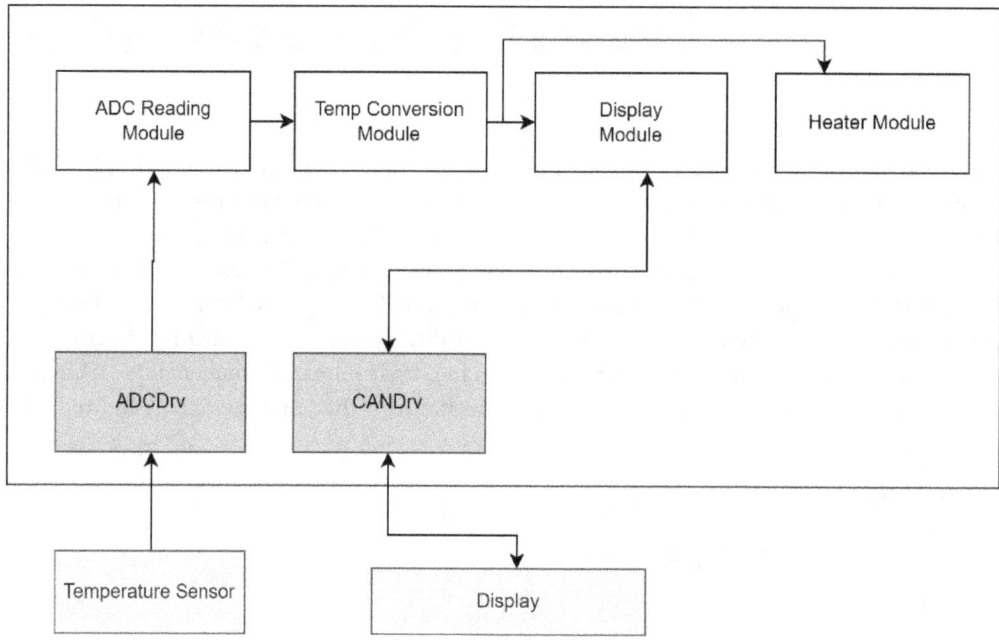

Figure 5.1 – Example of temperature sensor

The abstracted ECU software might utilize the ADC peripheral to acquire the signal value (Temperature). This process is likely handled by the ADC reading software module. The ADC values are expected to undergo a filtration process, and the filtered value will be forwarded to a temperature conversion SW module. This specific SW module is responsible for converting the filtered ADC value into a physical temperature value, which will eventually be presented on the display Module.

Let's get acquainted with the different SW modules and their roles:

- **ADC reading module**: This reads the ADC value, applies conversion, and provides it to a temperature conversion SWC. The following code snippet illustrates how a system is implemented outside of the AUTOSAR context. It includes an `analog_to_digital` function that converts analog signals to digital by reading a specific ADC channel. The `adc_task_10ms` task retrieves analog values from a temperature sensor, converts them into digital values, and sends them to a temperature processing module through `TCM_ProcessTemperature`:

    ```c
    #define ADC_RESOLUTION 1024   // Assuming a 10-bit ADC
    #define TEMP_SEN_CHANNEL 2
    int analog_to_digital(int  channel) {
        // Convert analog value to digital
        int digital_value = (int)(rREG_ADC[channel].rawval * ADC_RESOLUTION);
        return digital_value;
    }
    void adc_task_10ms() {

        int digital_value=analog_to_digital(TEMP_SEN_CHANNEL);
          TCM_ProcessTemperature(digital_value);
    }
    ```

- **Temp conversion module**: This takes in the digital value, applies calibration and conversion, and prepares it for display. When there is a temperature change, it sends a request to the display module to show the updated value on the car's dashboard. The code defines a function for processing temperature values obtained from the ADC Reading Module, applies calibration if needed, and sends updated temperature values to a display module. This process avoids redundant updates if the temperature remains unchanged:

    ```c
    #define CALIBRATION_FACTOR 1.0
    #define TEMPERATURE_CONVERSION_FACTOR 0.01
    float current_temperature=0;

    std_return  TCM_ProcessTemperature (int digital_value) {
        float calibrated_value = digital_value * CALIBRATION_FACTOR;
        float temperature = calibrated_value * TEMPERATURE_CONVERSION_FACTOR;
        current_temperature = temperature;
        return(E_OK);
    }
    void processing_task_10ms() {
        static float previous_temperature = -1;
        Static T_TState current_state= RUNNING_STATE;
        If .... some code
    ```

```
        If ((current_temperature != previous_temperature) &&
someothercondition()) {
            DIS_UpdateTemp(temperature , &current_state);
            previous_temperature = temperature;

        }
    }
```

- **Display SW module**: This displays the temperature value on the car's dashboard. The following code showcases functionalities for displaying temperature readings. The `display_task_50ms` task runs every 50 ms, retrieves current temperature values, and sends them as a CAN message to the display unit if some conditions were satisfied:

```
float updated_temperature=0;
void DIS_UpdateTemp(float temperature, unsigned char* state) {
     updated_temperature= temperature;
     /*some code to use state*/
     ............
}

void display_temperature(float temperature) {
     uint8_t *temp_ptr = (uint8_t*)&temperature;
     Uint8 data[8] =0;
     data[0] = temp_ptr[0]
     ......

     if (getsomestate() == TRUE) {
           SendSignalOverCAN(0x7A1,&data);
     }
}
void display_task_50ms() {
     If(somecondition == TRUE) {
display_temperature(temperature);
     }
}
```

- **CANDrv**: This is a low-level driver CAN implementation to send a message to the physical display module.
- **ADCDrv**: This is a low-level driver ADC implementation to configure and read ADC channels

Figure 5.2 shows how different software modules interact to read, process, and display temperature data in an automotive system.

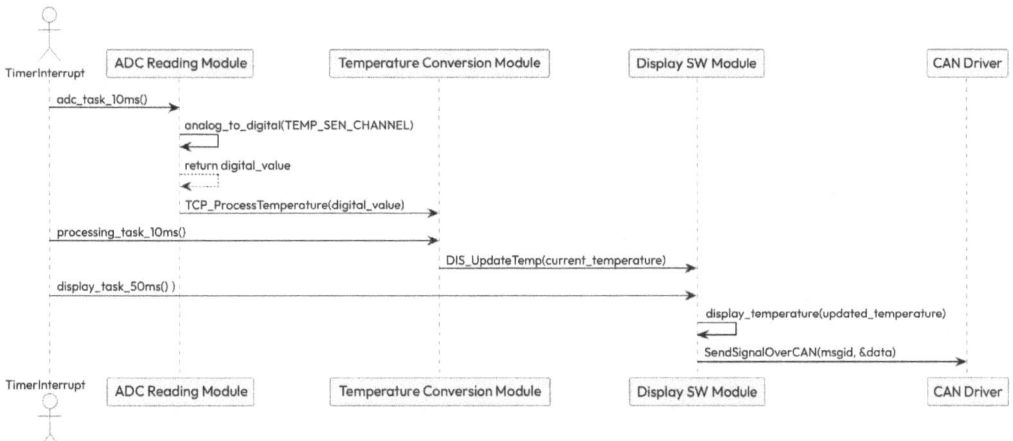

Figure 5.2 – Sequence diagram showing Temp Sensor module interaction

Information flows can be categorized based on their nature. Like outputs from an analog to a digital converter, streams of values might use **sender-receiver (SR) interfaces**. For behavior requests and their responses, **client-server interfaces** are employed. These are the primary types, and we will discuss them in the following sections. Other information flow methods, such as the mode-switching and parameter interfaces, are outside the scope of this discussion.

> Note
>
> Please take a moment to thoroughly review the example provided. While it's a simple illustration, it's important for our upcoming exploration of AUTOSAR interfaces. This example will serve as a practical reference point, helping to understand the concepts we'll discuss. Ensuring a clear understanding now will make the subsequent sections much more accessible and meaningful. The same example will be used again to re-design it following AUTOSAR interfaces and methodology.

Sender-receiver communication

SR interfaces manage data exchange between SWCs in AUTOSAR. These interfaces support various communication patterns, including the 1-to-N model where one sender transmits data to multiple receivers, and the N-to-N model where multiple senders can transmit data to multiple receivers. There are two types:

- **Sender (P Port)**: Sends out information; the interface is referenced by a providing port
- **Receiver (R Port)**: Awaits incoming information; the interface is referenced by a required port

For example, *Figure 5.3* demonstrates a 1-to-N configuration, an SWC responsible for calculating vehicle speed (sender) can send this data to multiple other SWCs (receivers). This data exchange can be periodic or event-driven.

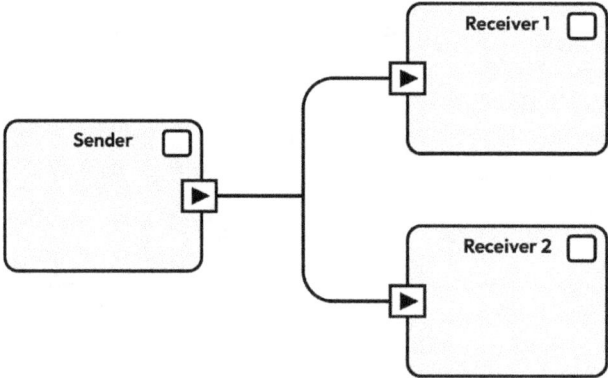

Figure 5.3 – SR 1-to-N example

Within the context of an SR interface, **data elements** are the fundamental units of information exchange. You can think of a data element as a virtual wire transmitting information such as vehicle speed, yaw rate, acceleration, and other data between SWCs. While it's common to see a single data element being transmitted, the SR interface is flexible enough to accommodate multiple data elements, allowing for a richer and more comprehensive data exchange.

> **Note on data elements**
>
> For simplicity, you can think of a data element in an SR interface as a global variable that is accessible to multiple SWCs through a defined interface. A group of data elements can be viewed as a structured set of data shared among these components.

For instance, consider a vehicle's onboard system that measures speed. The speed data, once captured, can be sent to multiple components:

- One component might use it to display the current speed on the dashboard
- Another might use it to adjust the vehicle's adaptive cruise control
- A third might log the data for diagnostic purposes

This multicast nature of the SR interface ensures that all these components receive the speed data simultaneously and can act upon it in almost real time. By now, you have grasped that software can be described in an XML format to define the software configuration, interfaces and behaviour for software units.

Let's explore how the SR interface is modeled in AUTOSAR XML (ARXML).

AUTOSAR XML (ARXML) modeling

In this section, we explore the structure and syntax of AUTOSAR XML (ARXML) files, providing examples to help readers understand how data elements and interfaces are modeled in AUTOSAR systems.

> **Note**
> The ARXML code that follows is provided as an example for illustration purposes and to help you become familiar with the ARXML syntax. It may not be entirely accurate and could differ slightly depending on the specific AUTOSAR schema.

This ARXML snippet defines a specific SR interface named `If_SR_SenderReceiverTest`, which is intended to handle the sending or receiving of the `speedVehicle` data element of type `uint32`.

```xml
<SENDER-RECEIVER-INTERFACE>
    <SENDER-RECEIVER-INTERFACE UUID="2DABF647-F234-404A-9669-58A29FABF1BC">
        <SHORT-NAME>If_SR_SenderReceiverTest</SHORT-NAME>
        <IS-SERVICE>false</IS-SERVICE>
        <DATA-ELEMENTS>
           <VARIABLE-DATA-PROTOTYPE UUID="D1368B17-1925-4F77-B4A8-520A992CED1D">
              <SHORT-NAME>VehicleSpeed</SHORT-NAME>
              <SW-DATA-DEF-PROPS>
                <SW-DATA-DEF-PROPS-VARIANTS>
                  <SW-DATA-DEF-PROPS-CONDITIONAL>
                    <SW-CALIBRATION-ACCESS>NOT-ACCESSIBLE</SW-CALIBRATION-ACCESS>
                  </SW-DATA-DEF-PROPS-CONDITIONAL>
                </SW-DATA-DEF-PROPS-VARIANTS>
              </SW-DATA-DEF-PROPS>
              <TYPE-TREF DEST="IMPLEMENTATION-DATA-TYPE">/AUTOSAR_Platform/ImplementationDataTypes/uint32</TYPE-TREF>
           </VARIABLE-DATA-PROTOTYPE>
        </DATA-ELEMENTS>
     </SENDER-RECEIVER-INTERFACE>
```

As a developer, you won't typically write ARXML code directly, but having a basic understanding of the format is helpful for debugging implementations. Here is an explanation:

- `<SENDER-RECEIVER-INTERFACE>`: This tag marks the beginning of the SR interface definition.
- `UUID`: This unique identifier helps distinguish this interface from others within the project.
- `SHORT-NAME`: This provides a user-friendly name for the interface, making it easier to understand its purpose.

- **IS-SERVICE**: This attribute indicates whether the interface offers a service (typically true for service-oriented communication) or simply facilitates data exchange (usually false for SR). In this example, it's set to `false`, signifying data exchange.
- **<DATA-ELEMENTS>**: This section defines the data that can be transmitted through the interface.
- **<VARIABLE-DATA-PROTOTYPE>**: This tag defines a specific variable within the interface.
- **UUID**: Like the UUID interface, each variable data prototype has its own unique identifier.
- **SHORT-NAME**: This provides a name for the data variable, making it clear what type of data is being exchanged.
- **SW-DATA-DEF-PROPS**: This section can contain various properties related to the data definition, but in this example, it focuses on calibration access.
- **<SW-DATA-DEF-PROPS-VARIANTS>**: This section allows for defining different variants or access rights for the data under specific conditions.
- **<SW-DATA-DEF-PROPS-CONDITIONAL>**: Here, the SW-CALIBRATION-ACCESS is set to NOT-ACCESSIBLE, indicating that the data cannot be modified through calibration tools. This might be appropriate for read-only sensor data.
- **TYPE-TREF**: This element specifies the data type of the variable. Here, it references a data type named `uint32` located under the `/AUTOSAR_Platform/ImplementationDataTypes/` path. This suggests the data being exchanged is a 32-bit unsigned integer.

Building on the previous example, we have seen how an SR interface is modeled using ARXML. Now, let's explore how a sender (provider) port is modeled and how it references the SR interface we discussed earlier.

Provided Port as Sender

This ARXML snippet defines a port prototype named in `Example_SenderPort`, which uses a non-queued sender communication specification to transmit the `VehicleSpeed` data element through the SR interface located at `/PortInterfaces/If_SR_SenderReceiverTest/`:

```
<P-PORT-PROTOTYPE>
    <SHORT-NAME>Example_SenderPort</SHORT-NAME>
    <PROVIDED-COM-SPECS>
        <NONQUEUED-SENDER-COM-SPEC>
            <DATA-ELEMENT-REF DEST="VARIABLE-DATA-PROTOTYPE">/PortInterfaces/If_SR_SenderReceiverTest/VehicleSpeed </DATA-ELEMENT-REF>
        </NONQUEUED-SENDER-COM-SPEC>
    </PROVIDED-COM-SPECS>
    <PROVIDED-INTERFACE-TREF DEST="SENDER-RECEIVER-INTERFACE">/PortInterfaces/ If_SR_SenderReceiverTest/ </PROVIDED-INTERFACE-TREF>
</P-PORT-PROTOTYPE>
```

The required port ARXML syntax example would be quite similar to the provided port. However, instead of <P-PORT-PROTOTYPE>, it would use <R-PORT-PROTOTYPE>, and the <NONQUEUED-SENDER-COM-SPEC> communication specification would be replaced with <NONQUEUED-RECEIVER-COM-SPEC>. In exploring the ARXML Schema, you will encounter different options and properties for the communication channel that provide more flexibility. For example, you might find options for queued communication specifications such as <QUEUED-RECEIVER-COM-SPEC>, which can handle buffered communication. The required port ARXML syntax example would be quite similar to the provided port. Perhaps, rather than P-Port, it would be R-Port and a different property for the <NONQUEUED-SENDER-COM-SPEC> communication channel would be <NONQUEUED-RECEIVER-COM-SPEC>. To understand the syntax more deeply, you need to refer to the XMLSchema for AUTOSAR.

> **Note**
> An XML schema defines the structure, content, and constraints of XML documents. In AUTOSAR, an ARXML (AUTOSAR XML) schema follows the general principles of XML schemas but includes specific rules and structures to model automotive software components and interfaces.

In the next section, we will explore the different communication types available for the SR interface within the AUTOSAR framework. By communication types, we mean the methods by which data is exchanged between SWCs using the SR model.

Communication types

The SR interface can be utilized in two different ways: explicit and implicit communication mechanisms. Let's explore the properties of each mechanism and why it might be preferred to use one over the other.

Implicit SR communication

There are two types of this communication model:

- IMPLICIT_SEND: When a runnable uses IMPLICIT_SEND to publish a data element, once the runnable concludes, the **run-time environment** (**RTE**) ensures that the most recent value of the data element becomes accessible to the data-element receivers.
- IMPLICIT_RECEIVE: Typically, when a runnable intends to read a data element, it does so by explicitly calling the RTE. However, with IMPLICIT_RECEIVE, the runnable directly accesses the value of the data element that was available at the start of its execution. The transmitted is stored in a buffer, so its value remains unchanged during the lifetime of the runnable. The following sequence diagram, *Figure 5.1*, illustrates the behavior of the implicit SR interface, where SWC B publishes data and SWC A receives the data.

Note

Runnable B has a higher priority than Runnable A. Thus, Runnable B can interrupt Runnable A and modify/read the data. We will be introduced to the concept of priority and tasks when we explore the AUTOSAR operating system within *Chapter 6*.

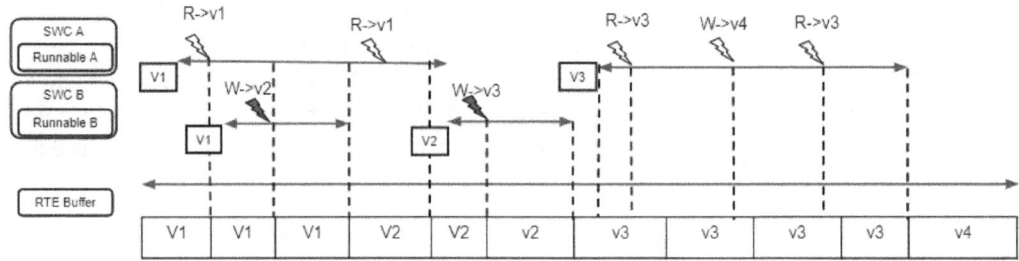

Figure 5.4 – Implicit write sequence diagram

Do you recall seeing the RTE described as the glue layer for communication? Next, we'll explore the generation of RTE APIs. Understanding how these APIs are created and managed will help you effectively debug, understand, and implement them.

How RTE APIs are generated

Here is the syntax for RTE APIs for the sender and receiver interfaces:

```
Rte_IRead : DataElementType  Rte_
IRead_<RunnableName>_<PortName>_<DataItemName>()
Rte_IWrite : Rte_
IWrite_<RunnableName>_<PortName>_<DataItem>(DataElemementType Data)
```

The following details explain the components of the RTE API syntax used for sender and receiver interfaces:

- `DataElementType`: The type of data being read.
- `Rte_IRead`: The prefix indicates it's an RTE read function.
- `<RunnableName>`: The name of the runnable where this read operation is performed.
- `<PortName>`: The name of the port from which data is read through.
- `<DataItemName>`: The specific data item being accessed.

Let's look at an example. If you have a runnable named `ReadSensorData`, a port named `SensorPort`, and a data item named `Temperature`, the function will look like `float Rte_IRead_ReadSensorData_SensorPort_Temperature()`.

Use case for Implicit Read

Consider a control algorithm that requires consistent temperature data throughout its execution. When the algorithm (runnable) starts, it reads the initial temperature value. This value is buffered, and the control algorithm uses this stable data throughout its operation, ensuring reliable processing without mid-execution changes.

Explicit SR communication

There are two types of this communication model:

- EXPLICIT SEND: When a runnable uses EXPLICIT SEND, once data is updated, it will be published to all users at once, ensuring the last available data is always ready.
- EXPLICIT RECEIVE: Explicit calls are forwarded to RTE to retrieve the latest available data, if any. Transmitted values are not buffered; whenever the runnable reads the data elements, the most recent value is obtained.

Explicit sender/receiver communication can be queued or unqueued, but both methods result in the most recent data being published and read immediately. You can think of explicit communication as a global buffer within the RTE, where both senders and receivers have access to it when read and write operations are performed.

In the context of explicit communication, *Figure 5.5* shows the relation between two SWCs. SWC B utilizes the RTE to write a data element to a shared buffer. Concurrently, SWC A features a runnable that continuously reads from this buffer. The sequence diagram vividly illustrates the crucial relationship, where the data written by SWC B is precisely what SWC A's runnable reads.

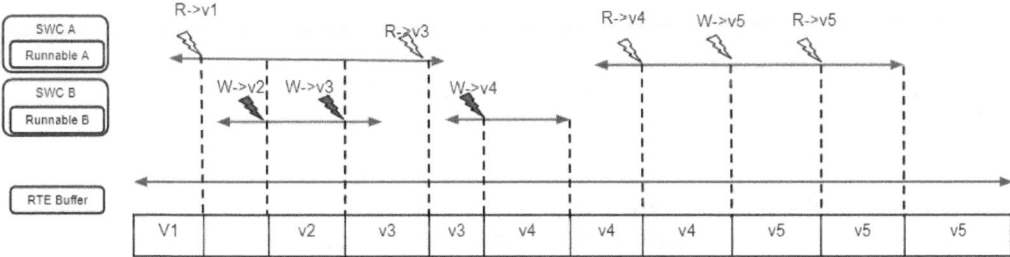

Figure 5.5 – Explicit unqueued write and read example

Figure 5.5 presents a similar example to the one discussed in *Figure 5.6*, but in this case, the example involves queuing. The data buffer in the RTE has a size of **3**. Pay attention to the relationship between reading from and writing to the buffer and observe how the queue behaves.

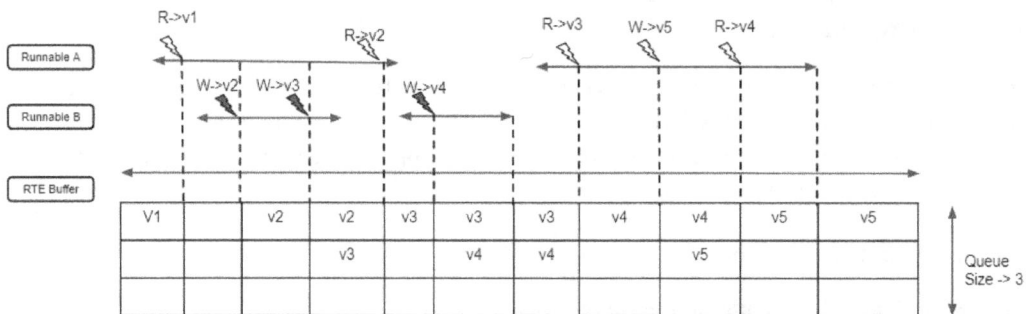

Figure 5.6 – Explicit queued send and receive example

Similarly to implicit SR communication, the RTE generates the APIs differently.

Here's how RTE APIs are generated:

```
Rte_Read  : DataElementType  Rte_Read_<Runnable>_<Port>_<DataItem>()
Rte_Write : Rte_Write_<Runnable>_<Port>_<DataItem>(DataElemementType Data)
```

The following details explain the components of the RTE API syntax used for explicit sender and receiver interfaces:

- `Rte_Write`: The prefix indicating that this is an RTE write function.
- `<Runnable>`: The name of the runnable that is performing the write operation.
- `<Port>`: The name of the port where the data will be written.
- `<DataItem>`: The specific data item being updated.
- `DataElementType Data`: The data being written, with its type specified.

> **Note**
> When it comes to the unqueued variant of the Explicit sender-receiver, it is `Rte_Write` for sending and `Rte_Read` for receiving. For the queued variant, it's `Rte_Send` and `Rte_Receive`, respectively.

Use case for Explicit read

Consider a system that monitors battery voltage. The monitoring system needs to check the battery voltage at specific intervals to ensure it remains within safe limits. When the monitoring system (runnable) initiates a voltage check, it explicitly requests the latest battery voltage through an RTE API call. This allows the system to obtain the most current data exactly when needed.

After exploring the SR interface, which primarily deals with periodic data streams, it's essential to recognize that not all interactions between SWCs are based on continuous data flows. There are scenarios wherein a component might need a distinct service or action from another. This is where the need for the *client-server interface* arises, which we will discuss next.

Client-server communication

Often abbreviated as **C/S interfaces**, this facilitates service-oriented communication between SWCs in AUTOSAR. These interfaces support a request-response communication pattern, whereby a client sends a service request to a server, which processes the request and returns a response. For example, in a client-server configuration, a SWC responsible for controlling vehicle lights (client) can request the status of the lights or send a command to turn them on or off. The server component processes these requests and provides the necessary responses. This interaction can be synchronous, where the client waits for the server's response, or asynchronous, where the client continues processing and handles the response later.

This is the client-server model, which is a fundamental concept in software design. Here, one SWC (the *server*) offers functionalities for another SWC (the *client*) to utilize:

- **Client (R Port)**: This is the component that *initiates* a request for a service. It acts as the consumer in the interaction.
- **Server (P Port)**: This component *responds* to the client's request by providing the desired services. It acts as the provider in the interaction.

This model is structured in a service-oriented manner, emphasizing the interaction between the client and server. Operations within the client-server interface are the specific functionalities or services that the server provides. These operations can have arguments, which are essentially the data or parameters required for the operation to execute. Arguments can be categorized based on their direction:

- **Set (in)**: These are input arguments. They provide the necessary data to the server. For example, in a calculator operation to add two numbers, the numbers themselves are *in* arguments.
- **Get (out)**: These are output arguments. The server returns these after processing the input. Using the calculator example, the sum of the two numbers would be an *out* argument.
- **Bi-directional (in-out)**: These arguments can both accept input and provide output. They are useful in scenarios where data needs to be both received and sent back after processing.

Figure 5.7 shows a client-server interface interaction, wherein two SWCs are exchanging information through calling operations or services provided by a SWC:

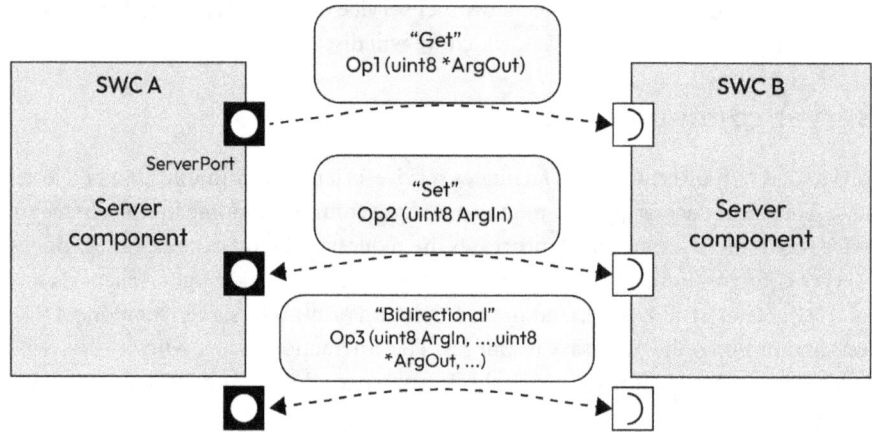

Figure 5.7 – Client-server interface

Let's explore how a client-server interface is modeled in AUTOSAR.

ARXML modeling

Let's take a look at the ARXML modeling of client-server interfaces in AUTOSAR. This ARXML snippet defines a specific client-server interface named `If_CS_LightControl`, which is responsible for controlling vehicle lights by providing operations to turn the lights on or off. This will provide a visual and conceptual understanding of the modeling of the C/S interface.

The following ARXML snippet defines a specific client-server interface named `If_CS_ClientServerTest`. This interface is not marked as a service (`IS-SERVICE` is set to `false`). It includes a single operation named `RunService`, which takes two arguments: `Arg1` (an input argument of the `uint8` type) and `Arg2` (an output argument of the `uint8` type). This setup illustrates how client-server interactions are modeled, where the client requests the `RunService` operation, passing `Arg1`, and receives a response in `Arg2`:

```
<CLIENT-SERVER-INTERFACE>
<SHORT-NAME>If_CS_ClientServerTest</SHORT-NAME>
<IS-SERVICE>false</IS-SERVICE>
<OPERATIONS>
  <CLIENT-SERVER-OPERATION>
    <SHORT-NAME>RunService</SHORT-NAME>
    <ARGUMENTS>
        <ARGUMENT-DATA-PROTOTYPE>
      <SHORT-NAME>Arg1</SHORT-NAME>
      <TYPE-TREF DEST="IMPLEMENTATION-DATA-TYPE">/
```

```
AUTOSAR_Platform/ImplementationDataTypes/uint8</TYPE-TREF>
                <DIRECTION>IN</DIRECTION>
                <SERVER-ARGUMENT-IMPL-POLICY>USE-ARGUMENT-TYPE</
SERVER-ARGUMENT-IMPL-POLICY>
            </ARGUMENT-DATA-PROTOTYPE>
            <ARGUMENT-DATA-PROTOTYPE>
                <SHORT-NAME>Arg2</SHORT-NAME>
                <TYPE-TREF DEST="IMPLEMENTATION-DATA-TYPE">/
AUTOSAR_Platform/ImplementationDataTypes/uint8</TYPE-TREF>
                <DIRECTION>OUT</DIRECTION>
                <SERVER-ARGUMENT-IMPL-POLICY>USE-ARGUMENT-TYPE</
SERVER-ARGUMENT-IMPL-POLICY>
            </ARGUMENT-DATA-PROTOTYPE>
        </ARGUMENTS>
    </CLIENT-SERVER-OPERATION>
```

The following examples demonstrate how the ARXML syntax appears for both a service provider (server) and service consumer (client).

The ARXML snippet defines a provided port prototype for a client-server interface in an AUTOSAR SWC. The port, named `P_If_CS_ClientServerTest`, includes communication specifications under `PROVIDED-COM-SPECS`, detailing a server communication specification (`SERVER-COM-SPEC`) with an operation reference to `/SWCA/PortInterfaces/ If_CS_ClientServerTest /RunService` and a queue length of 1. Additionally, the PROVIDED-INTERFACE-TREF references the `/SWCA/PortInterfaces/If_CS_ClientServerTest` client-server interface:

```
<P-PORT-PROTOTYPE>
    <SHORT-NAME>P_If_CS_ClientServerTest</SHORT-NAME>
    <PROVIDED-COM-SPECS>
        <SERVER-COM-SPEC>
            <OPERATION-REF DEST="CLIENT-SERVER-OPERATION">/SWCA/
PortInterfaces/If_CS_ClientServerTest/RunService</OPERATION-REF>
            <QUEUE-LENGTH>1</QUEUE-LENGTH>
        </SERVER-COM-SPEC>
    </PROVIDED-COM-SPECS>
    <PROVIDED-INTERFACE-TREF DEST="CLIENT-SERVER-INTERFACE">/SWCA/
PortInterfaces/If_CS_ClientServerTest</PROVIDED-INTERFACE-TREF>
</P-PORT-PROTOTYPE>
```

Let's take a closer look at the communication modes in a client-server interface to understand their unique functionalities, how they operate, and the typical scenarios wherein they are most effectively applied.

Communication types

Communication between a client and a server can occur in two modes: synchronous or asynchronous, based on the timing and specific usage needs of the designed function.

Synchronous client-server communication

This refers to a communication method whereby the client sends a request to the server and waits for a response before proceeding further. In this mode, the client is blocked or waits until it receives a complete response from the server, which might involve a delay if the server takes time to process the request. The interaction is tightly coupled, and each request is followed by an immediate response before the client can continue with the next action:

- **Direct request and response**: In synchronous communication, when a client requests a service from the server, it waits for the server to process the request and send back a response before continuing its own processing.

- **Blocking nature**: The client is effectively *blocked* until the server completes the service and sends a response. This means the client cannot perform other tasks during this waiting period.

- **Immediate feedback**: The client immediately gets feedback from the server, whether it's the result of a computation, data retrieval, or acknowledgment.

- **Typical use**: Synchronous communication is suitable for scenarios wherein the client needs immediate results from the server to proceed. For example, a SWC might request a specific calculation and need the result before it can continue its operations.

As shown in *Figure 5.8*, the sequence diagram demonstrates a synchronous communication pattern whereby a client sends a request to a server, then waits for the server to complete its processing before receiving and acknowledging the response. This synchronous interaction ensures that the client and server operate in a coordinated and blocking manner, with the client actively waiting for the server's actions to complete before continuing its own tasks.

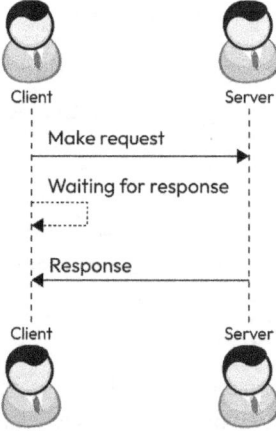

Figure 5.8 – Synchronous client-server call

Asynchronous client-server communication

This refers to a method whereby the client sends a request to the server but doesn't wait for an immediate response. Instead, the client continues its operation without halting for a reply from the server. The server processes the request separately and sends a response back when it's ready. This mode allows the client to perform other tasks or send additional requests without waiting for each individual response, leading to potentially increased efficiency and flexibility in handling multiple tasks concurrently:

- **Independent request and response**: In asynchronous communication, the client sends a service request to the server and then continues its own processing without waiting for a response. The server will process the request and send a response at a later point in time.
- **Non-blocking nature**: The client isn't blocked after sending a request. It can perform other tasks and handle the server's response whenever it arrives.
- **Delayed feedback**: The feedback from the server might be delayed, and the client typically uses mechanisms such as callbacks, events, or interrupts to handle the response when it's available.
- **Typical Use**: Asynchronous communication is ideal for scenarios wherein the client doesn't need immediate results from the server or when the server's processing might take an unpredictable amount of time. For instance, a SWC might request data from a server that requires extensive computation or retrieval from a slow storage medium.

As shown in *Figure 5.9*, the sequence diagram illustrates an asynchronous communication pattern whereby the client initiates a request to the server and continues its execution without waiting for the server to complete processing. The server processes the request independently and sends the response asynchronously. This asynchronous interaction allows the client and server to operate concurrently and perform tasks independently.

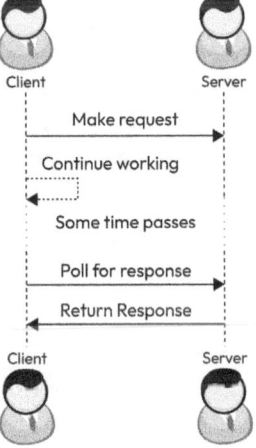

Figure 5.9 – Asynchronous client-server call

Now is the perfect time to revisit the example discussed at the beginning of this chapter in the *Explaining the information flow* section. We will use the same example to demonstrate how it is implemented in AUTOSAR.

Rethinking temperature sensor design

Let's revisit the example we initially mentioned in this chapter for temperature monitoring. We could rethink the example using interfaces and ports as follows:

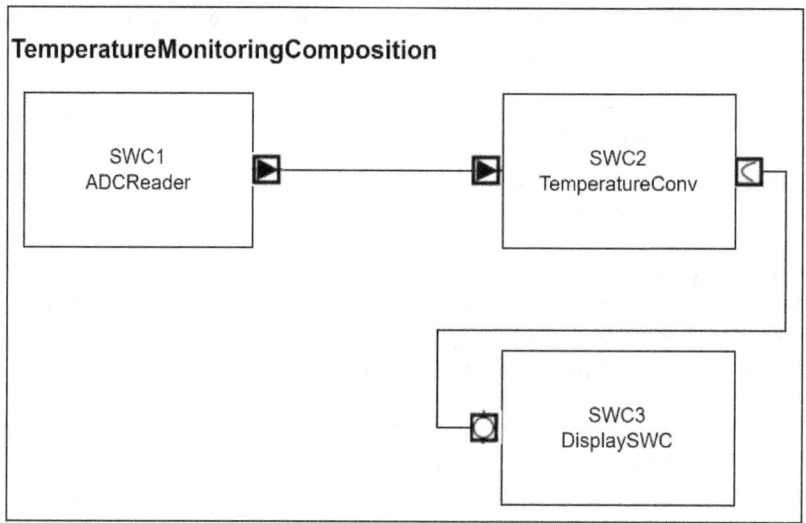

Figure 5.10 – Software composition for temperature monitoring SW

Figure 5.10 and *Table 5.1* illustrate a potential design of the SWCs and ports.

SWC	Port or interface	Type	Description
SWC1 - ADC Reader	PP_Temp	Provider port	Provides the digital temperature value to other SWCs
	Interface	Sender	Interface for sending the digital temperature value to the connected required port

SWC	Port or interface	Type	Description
SWC2 - Temperature Conversion	RP_Temp	Required port	Receive the digital temperature value from the connected Provider port
	Interface	Receiver	SR Interface for receiving a digital temperature value from the connected provider port
	RP_Display	Required port	Required Port to use the service to display the temperature
	Interface	Client-server	Interface for accepting display requests from the client port of the display module
SWC3 - Display SWC	PP_Display	Server port	Provides the service to accept display requests and display the temperature data
	Interface	Client-server	Interface for receiving display requests and displaying the temperature data from SWC2

Table 5.1 – Temperature monitoring example with AUTOSAR Interfaces and Ports.

Let's look at how the communication flow really takes place based on the proposed design:

1. SWC1 reads the temperature value and sends it to SWC2 using the SR interface.
2. SWC2 processes the temperature value and then requests SWC3 to display it using the client-server interface.
3. SWC3 receives the request and displays the temperature on the dashboard by using a sender-receiver interface to send a CAN message through the COM module.

You might be wondering when to select SR versus client-server interfaces, how to choose between Provide and Require ports, and how to decide between SR or client-server interfaces. These choices depend largely on the nature of the communication and the kind of interaction required.

When deciding between a `Provide` or `Require` port in AUTOSAR, consider whether the SWC is providing or consuming data or operations. A `Provide` port is used by the component that broadcasts data or offers a service, such as a sensor module sending ADC values or a module providing a temperature conversion service. Conversely, a `Require` port is used by the component that needs to consume data or request a service, such as a module receiving sensor data for calculations or a client requesting an algorithm execution. The choice is determined by whether the component is the source (`Provide`) or the consumer/requester (`Require`) of data or services.

SR is fitting for broadcasting data asynchronously, where a module needs to provide continual updates without knowing who the recipients are or expecting any return. For example, streaming sensor values to multiple components fits well here.

Client-server is used when a component must initiate an action or retrieve data from another component, often with the need for feedback or confirmation. This approach suits cases such as starting a process, executing an algorithm, or getting specific data where a response is necessary.

In some situations, it's possible to use either interface to share data depending on the requirements. However, an SR is typically used for ongoing data updates where responses are not required, while client-server is usually chosen when an interaction or command must be explicitly executed with an expected outcome.

> **Note**
> In this example, we have selectively focused on the ADC Reading Module, Temp Conversion Module, and Display Module, specifically illustrating the use of interfaces for providing ADC and temperature values, as well as display values. The detailed design of how the ADC Reading Module reads data from the driver is beyond the scope of this discussion. The goal is to give a glimpse of when to use SR interfaces, such as for continuous data transfer, and when to use client-server interfaces for direct service requests and responses.

As we transition from our discussion on ports and interfaces, it's essential to introduce the concept of *events* within the RTE.

Understanding events in AUTOSAR

An event can be looked at as an identifiable occurrence or state change in a system that may require a corresponding action or reaction. Events in AUTOSAR are specific occurrences or conditions that trigger certain actions or reactions within the system. They act as signals, informing SWCs that a particular condition has been met, and prompting them to execute specific functionalities.

Events not only facilitate synchronization and communication among SWCs but also significantly influence the system's overall performance and functionality. This section will explore the interdependent relationship between ports, interfaces, and events, highlighting their combined impact on the AUTOSAR ecosystem.

Various sources can generate events, such as external inputs, timers, or even other SWCs. They are typically handled by dedicated **event handlers** or **interrupt service routines**. *Figure 5.11* illustrates the sequence in which RTE events can be generated from various sources. These sources could include an OS alarm, a change in signals, or other triggers. The diagram demonstrates how the RTE processes and triggers the event, and finally, how the SWC executes the corresponding runnable.

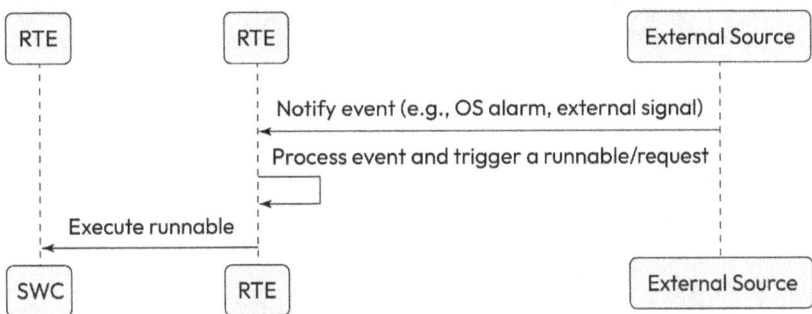

Figure 5.11 – RTE event generation and processing

To fully grasp the role of events in AUTOSAR, it's essential to understand how they interact with the ports and interfaces, which are the fundamental channels of communication between SWCs. Let's explore how these elements work together to ensure data flow and functionality within the AUTOSAR environment.

How do ports, interfaces, and events interact with each other?

Ports and interfaces define the pathways for information flow. They determine how different SWCs or modules communicate with each other and the external world. They also specify the data elements or operations that can be exchanged or invoked.

However, merely having these pathways is not enough. We need a mechanism to trigger actions based on the information received or changes in the system state. This is where events come into play. Events are triggers that cause certain actions to occur within the SWCs. They are associated with ports and interfaces to manage the flow of data or invocation of services.

For instance, an event can be generated when a sensor detects a specific condition such as a car door opening. This event then triggers a series of actions, such as turning on the interior lights. Therefore, an event is a trigger to an action, which is usually runnable in AUTOSAR, that defines (part of) the internal behavior of a component. The RTE controls this operation and requires a connection between your component and something else, which can be the OS, an SWC, and/or BSW modules.

Consider an abstracted scenario within an ECU wherein a temperature sensor's SWC detects a drastic temperature rise. The sensor SWC (SWC 1) periodically sends temperature data through its sender port using an SR interface. The main function in SWC 2, which is responsible for monitoring this data, performs periodic checks that are triggered by timing events. During one of these periodic checks, it detects the temperature rise. Upon detecting the critical temperature, SWC 2 immediately uses a client-server interface to request an emergency response from SWC 3. This request triggers an event in SWC 3, which then activates the cooling system without delay. This example, while simplified for clarity, illustrates how ports, interfaces, and events work together within an ECU to manage critical interactions between SWCs, ensuring prompt responses to important system changes.

You have now glimpsed how events play a central role in orchestrating the seamless functioning of the automotive ECU. To further understand this intricate system, let's delve deeper into events and get a closer look at their types.

Types of RTE events in AUTOSAR

The table provided offers a comprehensive overview of RTE events within the context of AUTOSAR. It presents a structured compilation of key RTE events, featuring columns detailing their abbreviations, event names, and comprehensive descriptions. The table aims to serve as a reference resource, providing a clear and concise breakdown of RTE events along with their respective functionalities and purposes.

Abbreviation	Event	Description
T	TimingEvent	Event based on a timer period
BG	BackgroundEvent	Recurring events perform background activities similarly to timer events but do not have a fixed period and have low priority
DR	DataReceivedEvent	Receive and process a signal received on an SR interface
DRE	DataReceiveErrorEvent	Triggered to collect the error status on the reception (SR Communication only)
DSC	DataSendCompletedEvent	Receive and process acknowledge transmission notifications (explicit SR communication only)
DWC	DataWriteCompletedEvent	Receive and process acknowledge transmission (implicit SR communication only)
OI	OperationInvokedEvent	Issued to run a server-side runnable; invoked by the client

Abbreviation	Event	Description
ASCR	AsynchronousServerCallReturnsEvent	Collect the result and status of an asynchronous CS operation
MS	SwcModeSwitchEvent	This event is raised upon a received mode change
MSA	ModeSwitchedAckEvent	Receive and process mode switched acknowledgment notifications
MME	SwcModeManagerErrorEvent	Triggered to react on errors occurred during mode handling
ETO	ExternalTriggerOccurredEvent	Triggered by the occurrence of an external event
I	InitEvent	Activate runnable for initialization purposes in case of RTE start or restart of a partition
THE	TransformerHardErrorEvent	Raised when during the transformation of received data, a hard transformer error occurs

Table 5.2 – RTE events

In this exploration, we will focus exclusively on three specific events within AUTOSAR: TimingEvent, OperationInvokedEvent, and DataReceivedEvent. These events have been chosen for their widespread use and significance in the AUTOSAR ecosystem:

- **TimingEvent**: This event stands out as a beacon of consistency and predictability. It serves as the metronome of the system, ensuring that specific functionalities are executed at regular intervals.

    Consider an automotive system where the **engine control unit** (ECU) needs to monitor the engine temperature every 10 milliseconds. In such a scenario, a TimingEvent can be created to trigger the temperature monitoring function of the ECU precisely at these intervals. This ensures the engine operates within safe temperature limits and any anomalies are detected and addressed in real time. Through this example, which we illustrate in the following figure, we can understand the role of TimingEvent events in maintaining the rhythm and reliability of automotive systems:

Designing and Implementing Events and Interfaces

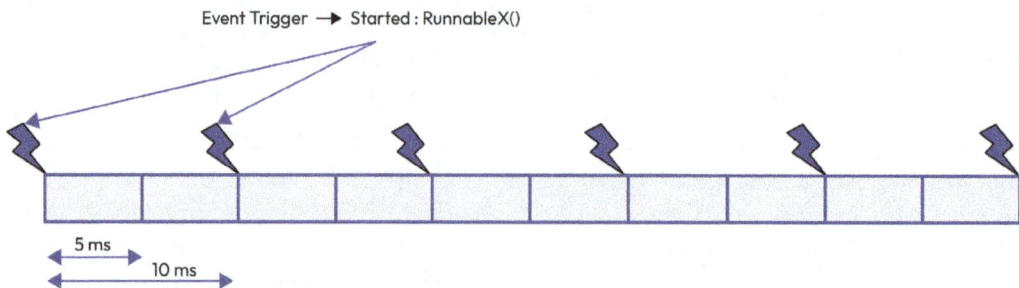

Figure 5.12 – TimingEvent in AUTOSAR

- **OperationInvokedEvent**: These events ensure that operations requested by clients are acted upon or initiated by the server provider. In essence, they facilitate the execution or initiation of requested operations by the server in response to client requests. Unlike events that operate on a set schedule, these events are all about immediacy, springing into action when a particular operation is invoked. They ensure that the system remains agile, as well as capable of responding to on-the-spot demands and executing functionalities based on real-time inputs.

Figure 5.13 illustrates how the Sensor SWC (SWC_B) requests a temperature calculation via the RTE, which schedules and triggers the `R_CalculateTemp` runnable in the Controller SWC (SWC_A), and then returns the calculated result back to the Sensor SWC.

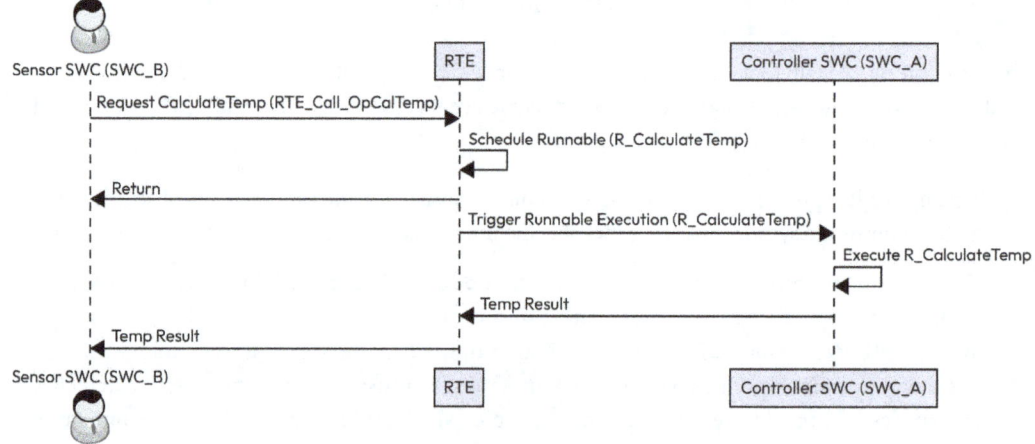

Figure 5.13 – OperationInvokedEvent sequence diagram

- **DataReceivedEvent**: This event plays a crucial role in ensuring the system reacts promptly to incoming data. This event is all about vigilance, and involves constantly monitoring specific data or signals and initiating actions as soon as they are detected. It ensures that the system remains alert and ready to respond to vital information, safeguarding both the vehicle's functionality and the safety of its occupants.

Envision a situation in a vehicle where a sudden collision is detected. The sensors send a signal to the airbag ECU, indicating the immediate need for deployment. A DataReceivedEvent event is triggered as soon as this signal is received, instructing the airbag to deploy without delay. This instantaneous response can make the difference between safety and injury, highlighting the importance of DataReceivedEvent events in ensuring timely reactions to critical situations. Through this example, we grasp the life-saving potential of such events in the automotive domain.

Figure 5.14 illustrates the sequence of utilizing data-received events. An SWC equipped with an SR interface publishes data to the RTE. Meanwhile, another SWC, labeled **SWC A**, acts as a user of the published data and desires a notification upon data reception to initiate a series of events. Therefore, **SWC B** writes data, and the RTE, recognizing **SWC A**'s need for notification, triggers an event to commence the execution of a runnable within **SWC A**.

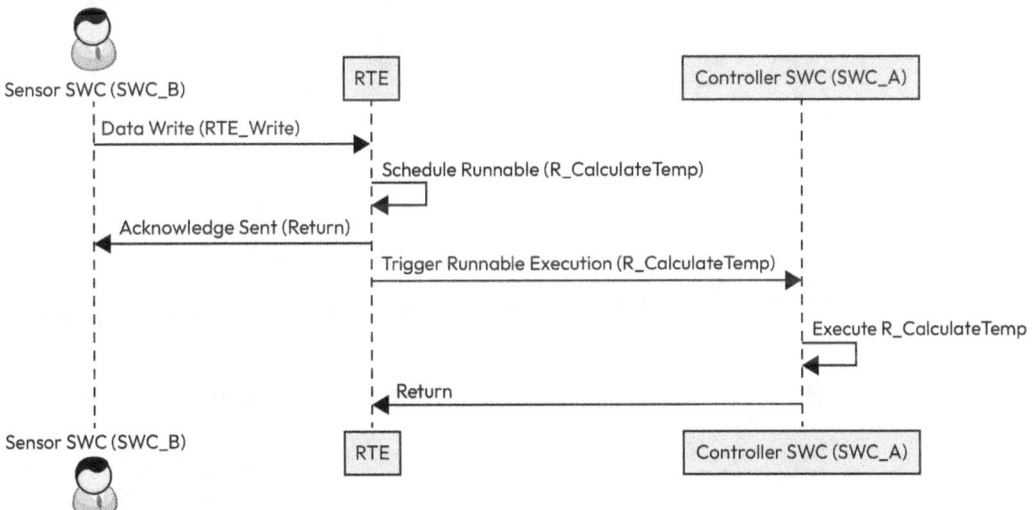

Figure 5.14 – DataReceivedEvent sequence diagram

> **Note**
>
> We have only discussed the most common events in this context. For a deeper understanding of how various events are used and what their specific applications are, please refer to the *RTE Events* chapter in the AUTOSAR documentation. Additionally, there are more details related to the events we have discussed, such as offsets for event scheduling and other advanced configurations. However, these advanced details are out of the scope of this discussion.

Now that we have discussed these events, let's revisit the temperature monitoring example and identify the types of events that would fit the design.

Events for temperature monitoring example

This table outlines the different SWCs, the types of events they use, the names of these events, and a brief description of their function within the temperature monitoring design.

SWC	Event type	Event name	Description
SWC 1 – ADC Reader	TimingEvent	TI_ReadADCValue	Triggers the ADC reading at regular intervals (every 10 ms) to read the digital sensor value
SWC2 – Temperature Conversion	DataReceiveEvent	DRE_ReceiveDigitalValue	Triggered when SWC 1 sends the digital temperature value to SWC 2
	TimingEvent	TI_ProcessTask10ms	Triggers the Temperature Calculation task every 10 ms
SWC3 – Display Module	OperationInvokedEvent	OIE_ProcessTemperature	Triggered when SWC 2 invoked the operation on SWC 3 to request an update on the temperature value
	TimingEvent	TI_DisplayTask50ms	Triggers the display task to send a CAN message every 50 ms

Table 5.3 – Event types and functions in a temperature monitoring system example

> **Note**
> This is just an example, and the specific events, ports, and operations used can vary depending on the design choices.

In this section, we explored the critical role of events in AUTOSAR and their interaction with ports and interfaces, providing a comprehensive design for seamless communication and synchronization among SWCs. We discussed key events such as OperationInvokedEvent events, which trigger runnables when specific operations are called; DataReceiveEvent events, which trigger runnables upon receiving new data; and TimingEvent events, which trigger runnables at regular intervals. These events, integrated with ports and interfaces, act like an engine, orchestrating the smooth operation of the AUTOSAR architecture.

Summary

This chapter unraveled the complexities of events and interfaces within the AUTOSAR framework. The narrative commenced by highlighting the role of events and interfaces in automotive systems, underscoring their importance in facilitating smooth data transitions, ensuring real-time responses, and upholding safety standards. As the chapter progressed, it delved deeper into advanced communication models, shedding light on both SR and client-server interfaces and further distinguishing between synchronous and asynchronous communication paradigms. An illustrative example of a car's temperature monitoring system served to contextualize these advanced features.

Now, we should be able to architect and establish robust communication pathways for automotive SWCs. Building upon the foundation laid in this chapter, the next chapter dives into the intricacies of the AUTOSAR operating system. We'll dissect its inner workings, exploring how it manages tasks, prioritizes execution, and interacts with hardware to ensure real-time performance. This in-depth exploration will equip you with the knowledge to design and implement robust automotive software solutions.

Questions

1. What is the primary role of events and interfaces within the AUTOSAR framework?
2. How do the SR and client-server interfaces differ in their communication models?
3. What is the significance of synchronous and asynchronous communication types in the AUTOSAR framework?
4. What are the different events provided by AUTOSAR?

Get This Book's PDF Version and Exclusive Extras

UNLOCK NOW

Scan the QR code (or go to `packtpub.com/unlock`). Search for this book by name, confirm the edition, and then follow the steps on the page.

Note: Keep your invoice handy. Purchases made directly from Packt don't require one.

6
Getting Started with the AUTOSAR Operating System

Building on our understanding of AUTOSAR and its multi-layered architecture, we are now prepared to examine more closely the pivotal role of the AUTOSAR **operating system** (**OS**) within this innovative framework and how it contributes to the overall process of automotive software development, from initial system design to performance tuning and fault management.

By exploring the role of the AUTOSAR OS and understanding its implications in automotive software development, readers will be equipped with a solid foundation to design, develop, and optimize software systems within the AUTOSAR framework. This chapter adds value by providing practical insights and a holistic understanding of the AUTOSAR OS and its contributions to the overall process of automotive software development. In this chapter, we focus on unraveling the complexity and functionality of the AUTOSAR OS by exploring the following main topics:

- A brief overview of the AUTOSAR OS
- Exploring the AUTOSAR OS architecture

A brief overview of the AUTOSAR OS

The core of the AUTOSAR platform is the AUTOSAR OS, which is responsible for efficiently managing the execution of software components on **electronic control units** (**ECUs**). With meticulous design, the AUTOSAR OS handles crucial functions such as task scheduling, resource allocation, event management, and alarm functionalities.

The AUTOSAR OS is positioned strategically within the AUTOSAR stack, interacting closely with all **basic software** (**BSW**) layers, managing and utilizing the **hardware** (**HW**) directly if needed. Let's take a look at the structure again:

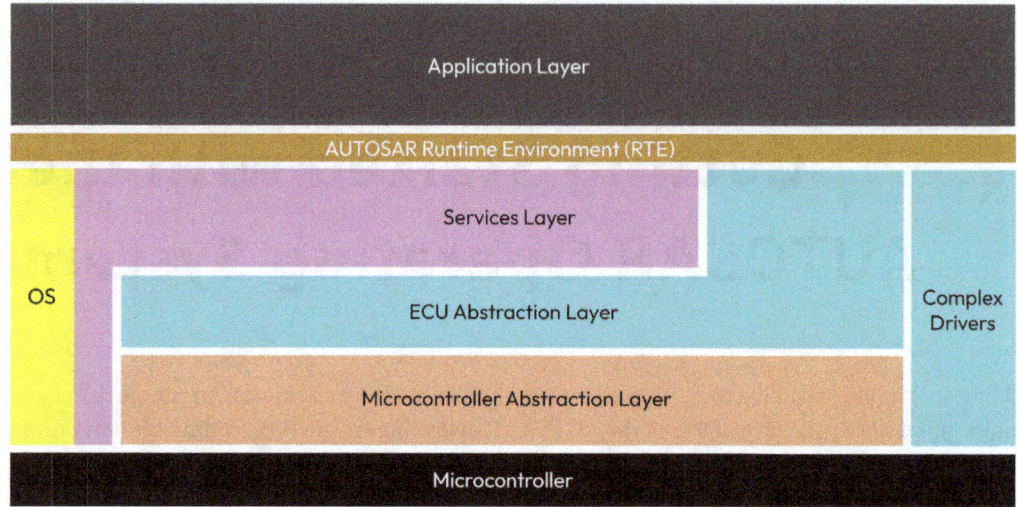

Figure 6.1 – The AUTOSAR OS in AUTOSAR architecure

The upcoming sections provide essential insights into the significance of **real-time operating systems** (**RTOS**) in embedded systems and the evolution of the **Offene Systeme und deren Schnittstellen für die Elektronik in Kraftfahrzeugen/Vehicle Distributed Executive** (**OSEK/VDX**) standard, which laid the foundation for the development of the AUTOSAR OS. Let's delve into these topics to understand the critical role of RTOS and the seamless integration of automotive software applications within the AUTOSAR framework.

Introduction to real-time operating systems

RTOS are specialized OS designed for embedded systems that require precise timing and deterministic behavior. Unlike general-purpose OS that prioritize maximizing throughput and user interactivity, an RTOS is engineered to provide predictability and ensure specific tasks are executed within strict time constraints. It excels at managing time and system resources accurately, delivering rapid response times to critical events.

The key difference between an RTOS and a regular OS lies in their primary focus. While a regular OS aims to provide a wide range of functionalities and services for diverse applications, an RTOS prioritizes deterministic behavior and real-time responsiveness. It guarantees that system tasks, particularly those with critical timing requirements, are executed within their designated time limits.

This unique characteristic makes RTOS crucial for systems where timing is critical, such as automotive control units. In the automotive industry, real-time performance requirements and absolute reliability are essential. Vehicles incorporate numerous ECUs that must seamlessly interact with one another, and safety-critical functions, such as how braking and engine control rely on precise timing and deterministic behavior. In the automotive industry, an RTOS is essential due to real-time performance requirements and absolute reliability. Modern vehicles incorporate numerous ECUs that must seamlessly interact with one another. In addition, safety-critical functions, such as braking, engine control, **advanced driver assistance systems** (**ADAS**), and others necessitate real-time operations to ensure safety and optimal performance. An RTOS, such as the AUTOSAR OS, provides the necessary features and services for these operations, including priority-based scheduling, fast interrupt processing, and inter-task communication and synchronization mechanisms and this has to be achieved within a small and deterministic memory footprint.

Now that we have explored the importance and unique characteristics of RTOS, let's explore the connection between RTOS and the OSEK/VDX standard, which played a significant role in establishing a consistent software architecture for ECUs in vehicles. We will explore how the foundational principles of the OSEK standard were adopted and further developed in the AUTOSAR OS to accommodate the evolving complexities of automotive software applications.

Introduction to OSEK

Emerging in the 1990s, the OSEK/VDX standard represented an automotive industry-led initiative. Its purpose was to create a consistent software architecture applicable across the multitude of ECUs deployed within a vehicle. With the advent of the AUTOSAR OS, the foundational principles of the OSEK standard were adopted and further developed to accommodate the evolving intricacies of automotive software applications.

Building on the groundwork laid by the OSEK OS standard, the AUTOSAR OS was designed for the particular needs of automotive control units. This included understanding and addressing automotive applications' unique constraints and real-time necessities. Essential features, such as preemptive and non-preemptive scheduling, were implemented for effective task management, promoting the optimal utilization of processing resources. In addition, strategies for resource management were integrated to mitigate challenges, such as **priority inversion**, a situation where lower-priority tasks might obstruct the completion of higher-priority ones. The AUTOSAR OS also incorporated mechanisms for event-driven task synchronization, delivering a more agile and efficient system for coordinating multiple tasks.

The AUTOSAR OS further enhances system performance through features such as support for multiple cores and flexible scheduling policies, enabling concurrent execution of tasks on distinct cores. It also upholds the principle of **stack monitoring**, a tool used to identify and prevent stack overflows, a frequent issue in real-time and embedded systems. Thus, the AUTOSAR OS builds on and extends the OSEK/VDX standard, effectively catering to the growing complexity and requirements of present and upcoming automotive software applications.

The following figure depicts the OS as the central engine of a computer system, illustrating its interactions with various modules and its relationship to driver layers:

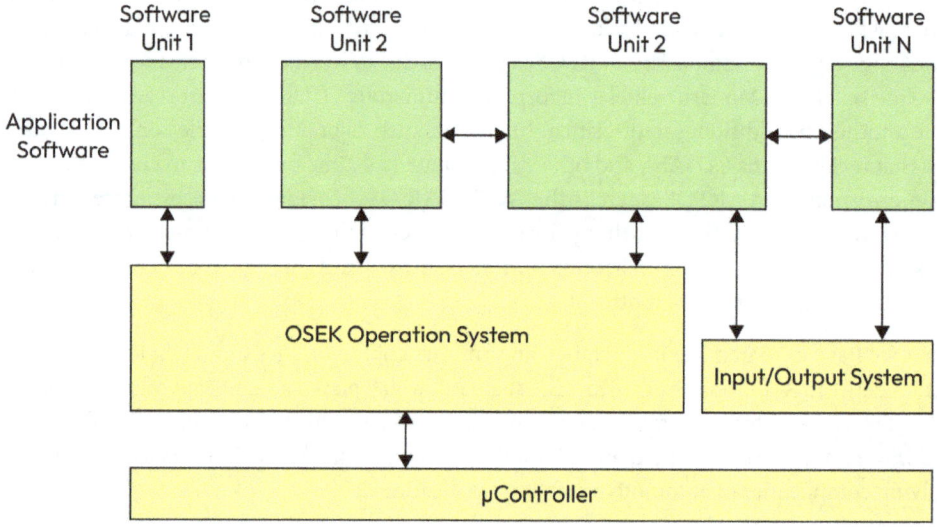

Figure 6.2 – Role of OS in an embedded software

The AUTOSAR OS configuration is a crucial step in the overall methodology, particularly during system integration. It involves configuring and integrating software components, the **run-time environment** (**RTE**), BSW modules, and the AUTOSAR OS itself.

Configuring the AUTOSAR OS

The configuration of the AUTOSAR OS happens during the **integration phase** of the AUTOSAR methodology. The configuration of the OS is crucial because it determines how it will manage system resources, handle interrupts, schedule tasks, and communicate with software components.

The AUTOSAR OS is closely related to the RTE, a middleware layer that enables communication between software components. These software components are typically developed independently and need a common way to communicate and exchange data. The RTE provides this capability and uses the services of the AUTOSAR OS to do so.

Software components, RTE, BSW modules, and the AUTOSAR OS are configured and integrated. The RTE generates a description file for each software component, outlining its interfaces and behavior. This information is used to generate the appropriate communication mechanisms (such as sending/receiving functions). The RTE also interacts with BSW modules, implementing low-level, hardware-dependent functionality, such as communication protocols and I/O management. BSW modules, as well as the OS, are configured based on the system requirements.

The following figure provides a visual representation of the OS configuration step within the AUTOSAR methodology, showcasing the process of configuring and integrating the OS and software components in an ECU:

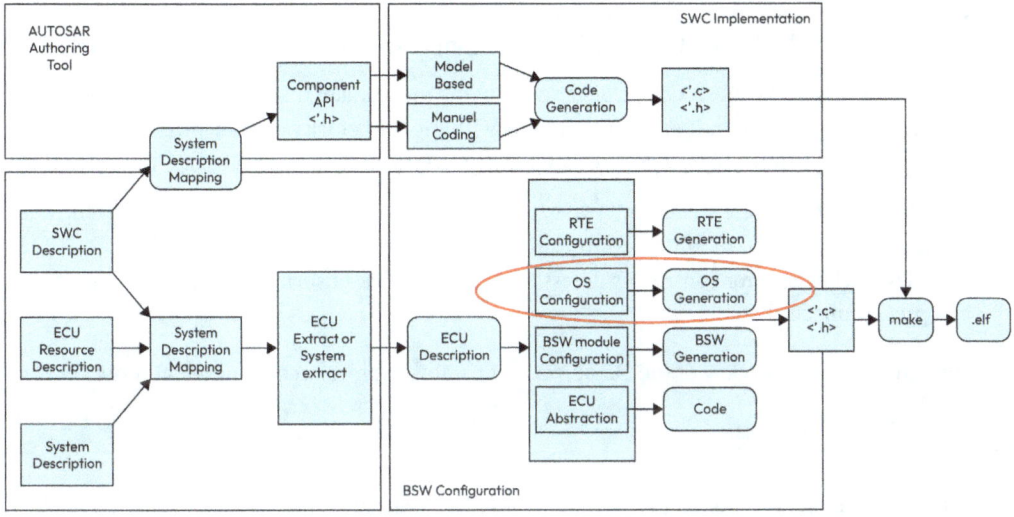

Figure 6.3 – Configuring the AUTOSAR OS

Now that you are familiar with what the AUTOSAR OS is, let's delve deeper into its architecture.

Exploring the AUTOSAR OS architecture

The AUTOSAR OS consists of several interconnected objects, each serving a specific purpose in managing the system resources and providing services to the software applications. These objects include the following:

- **Tasks**: In the AUTOSAR OS, tasks form the fundamental building blocks of execution. Essentially, these tasks are C functions invoked by the OS. They are distinct software elements that carry out specific functions and operate in a predetermined sequence, often activated by particular events or stimuli.

- **Alarms**: Alarms are time-based triggers that can be used to schedule the execution of tasks or activate events at specific points in time. They enable the AUTOSAR OS to implement time-driven behavior and meet strict timing requirements.

- **Events**: Events represent occurrences or conditions that can trigger the execution of tasks. Further, tasks, interruptions, or other system components can in turn generate events.
- **Services**: Services are pre-defined functions provided by the AUTOSAR OS to facilitate common operations, such as inter-task communication, memory management, and timing services. They offer a standardized interface for software components to access the OS functionality.
- **Scheduling**: The AUTOSAR OS includes a scheduling mechanism that determines the order and priority of task execution. It ensures that tasks with higher priority are executed before lower-priority tasks, enabling the system to meet timing requirements and guarantee real-time behavior.
- **Resources**: Resources are shared entities that multiple tasks can access and manage. They include hardware peripherals, memory, communication channels, and timers. The AUTOSAR OS provides mechanisms for controlling access to these resources, ensuring proper synchronization, and preventing conflicts.

Having provided a brief overview of the OS objects, let us now explore each component more deeply.

Tasks

The AUTOSAR OS offers two distinct task concepts for effectively subdividing and executing complex control software based on real-time requirements: **basic tasks** and **extended tasks**. By implementing tasks, the software can be conveniently divided into manageable parts, each with its framework for executing functions. In addition, the OS enables concurrent and asynchronous execution of these tasks, and the scheduler ensures an organized sequence of task execution.

Let's explore these two types of tasks in greater detail.

Basic tasks

Basic tasks only release the processor when certain conditions are met. These conditions include the following:

- The task terminating
- The OS switching to a higher-priority task
- The occurrence of an interrupt that triggers the processor to switch to an **interrupt service routine (ISR)**

Let's look at the **task state model** of a basic task. The task state model consists of a state machine that governs the behavior of tasks within the AUTOSAR framework. The following figure shows the state model for a basic task:

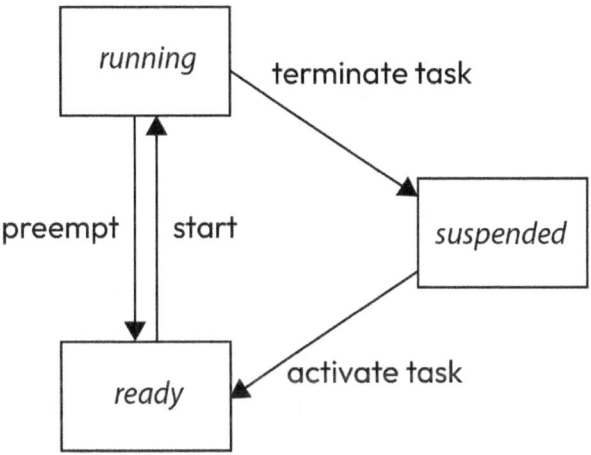

Figure 6.4 – Basic task state machine

Let's explore the components shown here:

- **Running**: When a task is running, the CPU is actively executing the instructions of that particular task within the embedded system. For instance, if you have an embedded system controlling the engine of a car, the task associated with the engine control module is actively running, continuously monitoring and controlling various engine parameters to ensure optimal performance and efficiency.
- **Suspended**: When a task is suspended, it is temporarily paused and not executed by the CPU. For example, this could happen when another task takes priority or when the task voluntarily yields control.
- **Ready**: A task in the ready state is prepared to be executed by the CPU but is currently waiting for its turn. The task is loaded into memory, and all its necessary resources are allocated.

To summarize the example, during normal driving, the engine management task remains running, continuously optimizing engine performance. However, if a potential collision is detected, the engine management task may be suspended temporarily to prioritize collision avoidance. Meanwhile, the braking task stays ready, monitoring inputs and waiting for the driver's command or an event that requires braking intervention.

It's important to note that in real automotive systems, numerous tasks may handle different functions simultaneously, and their states can dynamically change based on system requirements, priorities, and events.

Extended tasks

Extended tasks in the AUTOSAR framework provide added functionality compared to basic tasks. They have the capability to utilize the OS's `WaitEvent` call, enabling them to enter a waiting state. This waiting state allows the processor to be released and assigned to a lower-priority task without terminating the currently running extended task. However, the management of extended tasks is more complex for the OS, requiring additional system resources compared to basic tasks.

The states of an extended task are analogous to those of a basic task, with the inclusion of an additional state known as the *waiting* state, as shown in the following figure:

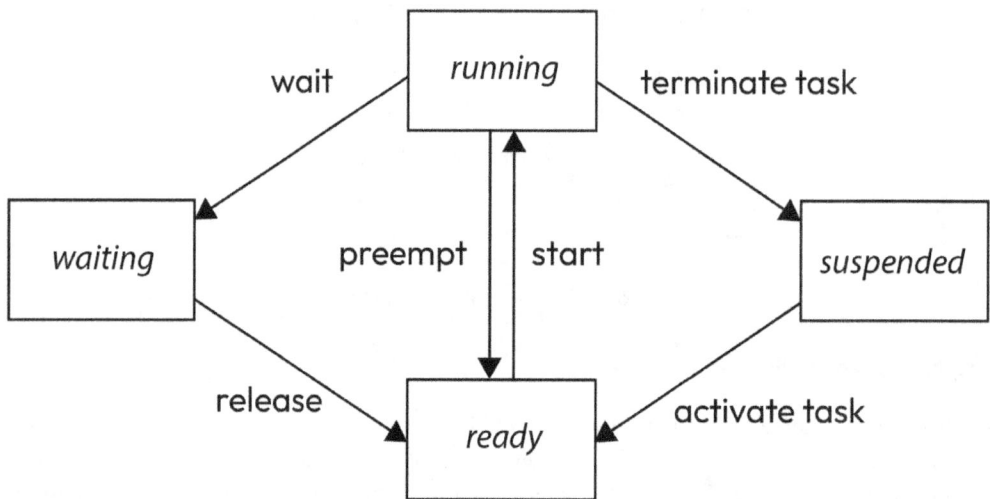

Figure 6.5 – Basic task state machine

The **waiting state** is an extended task-specific state in OS. In this state, a task is waiting for a particular event, condition, or resource before it can proceed with its execution. The task remains in the waiting state until the expected event or condition occurs, at which point it can transition to the running state to continue its operations.

In our example, in the embedded automotive domain, extended tasks can have additional states, such as **waiting** and **blocked**. The sensor data processing task runs continuously, the communication task may be suspended temporarily in case of errors, the diagnostics task waits for trigger events, the input handling task waits for user input, and a task can be blocked if it requires exclusive access to a shared resource. These extended states allow for more nuanced control and coordination of tasks within the embedded automotive system.

Interrupts play a crucial role in handling time-critical events and ensuring timely responsiveness within the automotive software system. In the next section, we will dive into the concept of interrupts, their types, and how they are managed within the AUTOSAR framework.

Interrupts

In embedded development, **interrupts** refer to signals or events that pause the normal execution of a program and temporarily transfer control to a specific section of code called an ISR. Interrupts are a fundamental mechanism for handling time-sensitive events and external interactions in embedded systems.

When an interrupt occurs, it interrupts the current execution of the program and diverts the processor's attention to the ISR associated with that particular interrupt. As a result, the ISR performs the necessary actions to respond to the event or signal, such as reading sensor data, servicing a hardware device, or handling an external input.

Interrupts are essential in embedded development, enabling efficient handling of time-sensitive events, responsiveness to external stimuli, and multitasking capabilities in embedded systems. They help ensure the system's stability, real-time performance, and effective utilization of system resources.

In the AUTOSAR OS, the functions for processing an interrupt are subdivided into two ISR categories:

- **ISR category 1**: ISRs in this category operate independently of OS services. Once the ISR completes its execution, processing resumes precisely from the instruction where the interrupt originated. This implies that the interrupt does not impact task management. Such ISRs have minimal overhead and operate efficiently within the system.
- **ISR category 2**: ISRs in this category interact with the OS. Once an ISR in category 2 finishes execution, the OS may perform a task switch or adjust task priorities. This is because category 2 ISRs can invoke OS services, such as activating tasks or setting events.

The following figure illustrates the presence of these two types of ISRs:

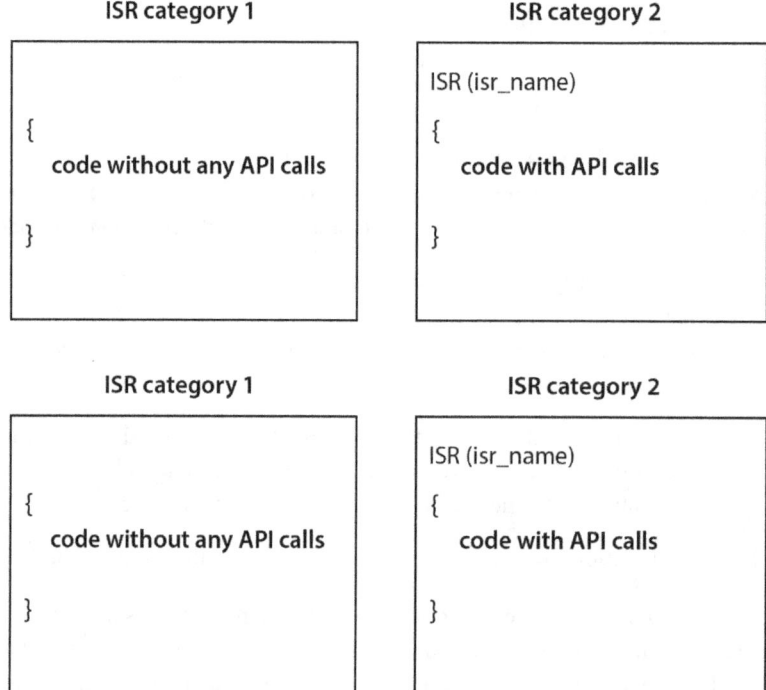

Figure 6.6 – ISR categories

In category 1 ISRs, they are not managed by the OS and operate independently. On the other hand, category 2 ISRs are managed by the OS through an OS frame. This differentiation allows for distinct handling and management of ISRs based on their type within the AUTOSAR framework.

Events

In the AUTOSAR OS, **events** are signaling mechanisms that provide vital cues to extended tasks about various state changes or happenings within the system. Essentially OS-managed entities, these events are closely linked to extended tasks, not as standalone elements but as auxiliary objects that complement these tasks. Each extended task is associated with a distinct set of events, effectively becoming the proprietor of these events. Events in AUTOSAR are characterized by two key attributes: the **owning task** and the **unique identifier**, often referred to as the *name* or *mask*. After activating an extended task, the OS systematically clears the related events, thereby readying them for their next assignment.

Events serve as conduits for binary communication, encapsulating various types of information, such as a timer's expiration, availability of a resource, and reception of a message, among other system-level occurrences. Therefore, the utility of events is pivotal in managing non-periodic situations that warrant an immediate response.

AUTOSAR's communication model is deeply rooted in this event-centric design, primarily aimed at fostering efficient task synchronization and communication in real-time environments. An event can be linked with multiple tasks, while a task can await several events. When a task is in a waiting state for a certain event, it remains blocked until the event is set, enabling precise coordination among other system-level occurrences. Therefore, the utility of events is pivotal in managing non-periodic situations that warrant an immediate response.

The AUTOSAR OS provides API functions for managing and manipulating events. These functions include the following:

- `SetEvent`: This function can be used to set an event for a particular task. When the event is set, the task waiting for this event will be unblocked and ready for execution.
- `ClearEvent`: This function is used to clear a specified event for a task.
- `WaitEvent`: This function blocks the execution of the calling task until one of its events is set.
- `GetEvent`: This function reads the currently pending events of a task.

Let's consider an example where we have two tasks, `Task1` and `Task2`. `Task1` sets an event (`Event1`), and `Task2` waits for this event to occur:

```
#define Event1 0x01
TASK(Task1)
{
    // ... Task1's processing ...
    // Set the Event1 for Task2
    SetEvent(Task2, Event1);
      // Terminate Task1
    TerminateTask();
}

TASK(Task2)
{
    // Wait for the Event1
    WaitEvent(Event1);
    // If the execution reaches this point, it means Event1 occurred
    // ... Task2's processing ...
    // Clear the Event1
    ClearEvent(Event1);
    // Terminate Task2
    TerminateTask();
}
```

In this example, Task2 uses the WaitEvent function to wait for Event1. Until Event1 is set, Task2 remains in the waiting state. Task1 uses SetEvent to set Event1 when it completes its processing. Once Event1 is set, Task2 moves from the waiting state to the ready state, and its execution continues. After handling the event, Task2 uses ClearEvent to clear Event1 and then terminates.

This mechanism allows tasks to synchronize their execution based on specific system states or occurrences, thus enabling efficient real-time system management.

The following figure provides a visual representation of events in the AUTOSAR framework, showcasing their setting, clearing, and sharing mechanisms among tasks or through the scheduler. It offers insights into how events are utilized to facilitate communication and synchronization between various tasks or executable entities:

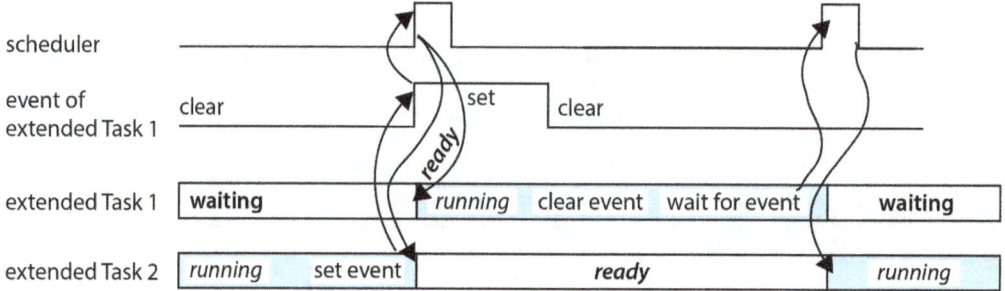

Figure 6.7 – Events between tasks

In the next section, we will explore the scheduling mechanisms employed by the AUTOSAR OS to ensure efficient task execution and resource allocation.

Scheduling

Scheduling in the AUTOSAR OS refers to the mechanism that determines the order and priority of task execution within the system. It plays a crucial role in managing the allocation of processor time and resources among different tasks, ensuring that the system meets its timing requirements and achieves real-time behavior.

The AUTOSAR OS supports different scheduling policies based on the specific needs of the system. Two common scheduling policies in AUTOSAR are as follows:

- **Full preemptive scheduling**: This type of scheduling enables a running task to be rescheduled at any instruction based on pre-defined trigger conditions set by the OS. With full preemptive scheduling, when a higher-priority task becomes ready for execution, the currently running task is placed in the ready state, and its execution context is saved. The higher-priority task is then scheduled and allowed to execute. This preemptive behavior ensures that tasks with higher priority are given immediate access to the processor, allowing critical tasks to meet their timing requirements. Once the higher-priority task completes its execution or becomes blocked, the previously preempted task can resume execution from the exact location where it was preempted.

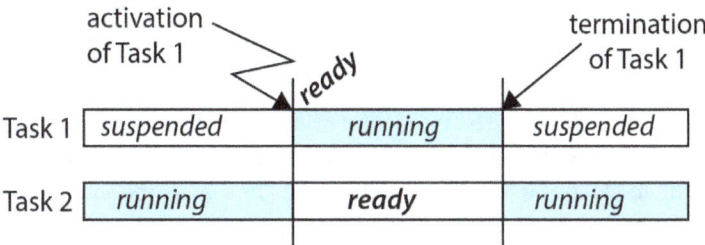

Figure 6.8 – Full preemptive scheduling

- **Non-preemptive scheduling**: The non-preemptive scheduling policy, also referred to as cooperative scheduling, enables tasks to continue execution until they willingly give up control or encounter explicit rescheduling points. In this policy, task switching occurs only through explicitly defined system services or specific rescheduling points. Under non-preemptive scheduling, a task retains control of the processor until it completes its execution, switches to a higher-priority task, or an interrupt occurs that triggers the execution of an ISR. However, non-preemptive scheduling imposes specific constraints on timing requirements. Notably, when a lower-priority task enters a non-preemptable section, it can cause a delay in the execution of higher-priority tasks until the next rescheduling point is reached.

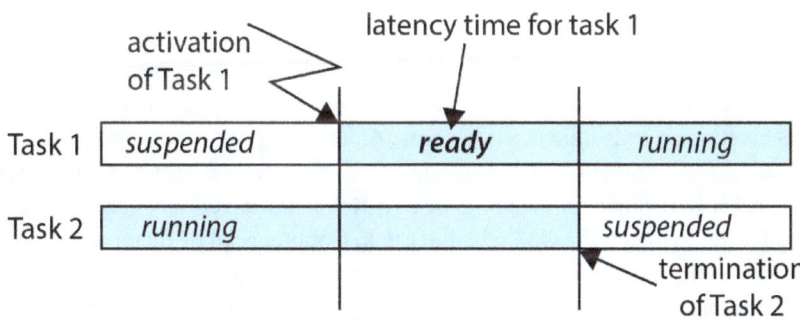

Figure 6.9 – Non-preemptive scheduling

Timing protection is ensured through the scheduling policy implemented in AUTOSAR, which incorporates a mechanism known as **highest priority first**. Under this policy, tasks with the same priority are grouped together and placed in a **first in, first out** (**FIFO**) stack. These tasks share a common resource. To maintain timing integrity, only one task within the group is allowed to occupy the shared resource at a time. When a task completes its operation and releases the resource, the next task in the FIFO stack can then acquire the released resource. This **sequential access** to the resource prevents multiple tasks from concurrently accessing it and ensures orderly execution.

Importantly, the scheduling policy based on the highest priority first does not allow one task to preempt another task within the same priority group. This means that tasks within a priority level are executed in a cooperative manner, without interrupting each other. These preemption rules, combined with the sequential access to shared resources, work together to provide timing protection for the system. By adhering to these rules, the system can effectively meet all its timing requirements, ensuring the timely execution of critical tasks and maintaining overall system reliability.

In the next section, we will delve into various task trigger mechanisms and their significance in facilitating efficient task execution, event handling, and inter-task communication.

Task trigger mechanisms

In AUTOSAR, task trigger mechanisms are used to activate tasks based on specific conditions or events. These mechanisms provide a way to initiate the execution of tasks in response to pre-defined triggers. There are two common task trigger mechanisms used in AUTOSAR: **alarms** and **schedule tables**. Let's explore these in detail in the following sections.

Alarms

The AUTOSAR OS provides functionalities for handling repetitive events. These events can include timers that generate interruptions at regular intervals, for example, triggering a timing event to the application or generating an activation event to a task. Alarms are pivotal time-driven entities in the AUTOSAR OS, designed to meet real-time systems' timing demands. They are intimately tied to **counters**, which are software timers that count in ticks. A **tick** can represent any time period, although, in practice, it's often set to match the OS's scheduler tick. Counters can be incremented in different ways, such as at fixed time intervals (system counter), via hardware interrupts (hardware counter), or manually by application software (software counter). Using counters provides a mechanism to handle time-based events. The relationship between counters and alarms is such that, when a counter hits a predetermined count value, the associated alarm expires and performs a pre-defined action. These actions primarily include the following:

- **Activation of a task**: Upon expiration, an alarm can activate a specific task. This allows for precise scheduling of tasks and implementing time-based behavior in AUTOSAR applications. For instance, a task could be activated every 100 milliseconds to check the status of a sensor.

- **Setting an event**: An alarm can also set an event for a particular task when it expires. This provides a way for tasks to synchronize their execution based on specific system states or occurrences.

The AUTOSAR OS oversees counters and associated alarms, ensuring that when a counter value reaches a specific threshold, the corresponding alarm triggers the configured action. This mechanism allows for efficient handling of timing requirements, resource optimization, and the implementation of time-dependent behavior in automotive applications.

The following is a pseudocode example demonstrating how an alarm can be used:

```
#define ALARM_CYCLE 10

TASK(Task1)
{
    // Task1 processing here
    // Terminate Task1
    TerminateTask();
}

void setup(void)
{
    // Start the alarm; it will expire after ALARM_CYCLE counter ticks,
and then every ALARM_CYCLE ticks
    SetRelAlarm(Alarm1, ALARM_CYCLE, ALARM_CYCLE);
}

void main(void)
{
    setup();
    while(1)
    {
        // Increment the counter
        IncrementCounter(Counter1);
        // Handle the alarms
        AlarmCheck();
        if (CheckAlarmExpired(Alarm1)) {
            // Activate Task1 when Alarm1 expires
            ActivateTask(Task1);
        }
    }
}
```

In this code, the `SetRelAlarm` function configures `Alarm1` to start after the `ALARM_CYCLE` counter ticks and to expire after every `ALARM_CYCLE` counter ticks. When the alarm expires, it triggers the activation of `Task1` in the main loop.

Next, let's discuss schedule tables.

Schedule tables

Schedule tables in the AUTOSAR OS are advanced timing mechanisms that provide a powerful way to define complex timing behaviors. They offer a higher level of abstraction than alarms and counters, allowing the precise scheduling of tasks and the setting of events based on time.

A schedule table can be thought of as a *timetable* of actions. It consists of multiple steps, each step corresponding to a point in time. When the time arrives for a specific step, the scheduled action (such as the activation of a task or the setting of an event) takes place.

The following figure showcases schedule tables and expiry points, highlighting their significance in orchestrating task execution and timing within the AUTOSAR framework. It emphasizes that each expiry point can possess an offset, enabling the activation of tasks and the setting of events at specific time instances:

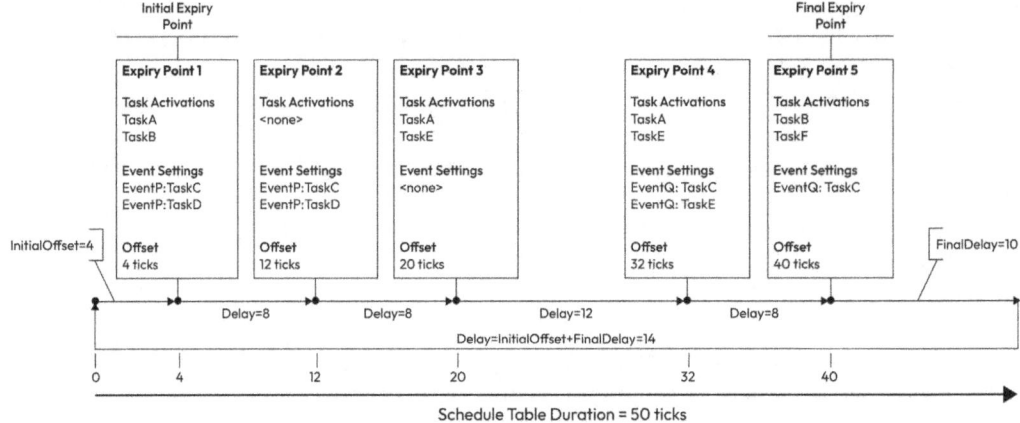

Figure 6.10 – Schedule table example

The AUTOSAR OS supports both relative and synchronized schedule tables. **Relative schedule tables** begin running as soon they are started, with the timing of each step relative to the start of the table. **Synchronized schedule tables**, on the other hand, are synchronized to an external time source.

Synchronized schedule tables play a vital role in inter-ECU synchronization. Multiple ECUs in a vehicle often need to coordinate their actions to achieve overall system objectives. For instance, different ECUs controlling various aspects of engine management (fuel injection, spark timing, etc.) need to act in a coordinated manner.

The following pseudocode is a simplified example of how schedule tables can be used for this purpose. Imagine we have two ECUs, ECU1 and ECU2, each running an instance of the AUTOSAR OS. ECU1 is the master, generating a synchronization signal every 100 milliseconds, and ECU2 is the slave, needing to perform certain actions synchronized to this signal:

```
// On ECU1 (master)
#define SYNC_CYCLE 100   // Sync signal every 100 ms

TASK(SyncTask)
{
    // Generate the sync signal
    SendSyncSignal();
    // Schedule the Task to run again in 100 ms
    SetRelAlarm(SyncAlarm, SYNC_CYCLE, SYNC_CYCLE);

    TerminateTask();
}

// On ECU2 (slave)

#define TABLE_START 0
#define TASK_STEP    50  // Task to be run 50 ms after receiving the
sync signal

TASK(SlaveTask)
{
    // Do something here...

    TerminateTask();
}

// When the sync signal is received...
void OnSyncSignalReceived(TickType syncTime)
{
    // Synchronize the schedule table to the received sync signal
    SyncScheduleTable(SlaveScheduleTable, syncTime);
}
```

In the preceding example, ECU1 generates a synchronization signal periodically using a task and an alarm. Upon receiving the sync signal on ECU2, the `SyncScheduleTable` function is used to synchronize the `SlaveScheduleTable` schedule table to the sync signal. This ensures that `SlaveTask` is activated 50 ms after receiving the sync signal, synchronizing the task execution on ECU2 to the actions on ECU1.

Through this, schedule tables in the AUTOSAR OS provide a powerful way to synchronize the operations of different ECUs, facilitating complex, time-coordinated behaviors across the whole vehicle system.

Now, let's shift our focus to OS resources; understanding how the OS manages and allocates resources is crucial for ensuring the efficient utilization of system capabilities. In the next section, we will explore the concept of OS resources and discuss their significance in the AUTOSAR framework.

OS resources

Resource management in the AUTOSAR OS plays an instrumental role in synchronizing shared resource access across tasks with varying priorities. These resources can encompass a broad spectrum, including scheduler management entities, sequential programs, memory regions, or specific hardware areas. The resource management system is a requisite across all conformance classes and can optionally be extended to include task and ISR coordination.

Let's now look at OS resources within AUTOSAR that represent critical elements, such as CPU power, memory, I/O devices, and communication channels:

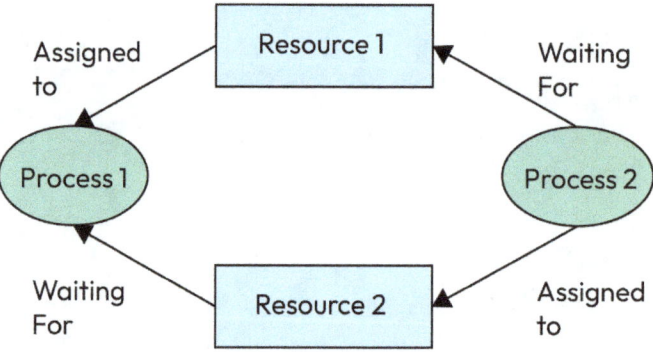

Figure 6.11 – Shared resource between two tasks

The core principles of resource management in the AUTOSAR OS are as follows:

- **Exclusive access**: The system ensures that two tasks cannot occupy the same resource simultaneously, preventing data inconsistency and corruption
- **Priority inversion prevention**: The AUTOSAR OS employs a priority inheritance protocol to prevent priority inversion – a situation where a higher-priority task is blocked, waiting for a lower-priority task holding a shared resource
- **Deadlock avoidance**: The OS ensures that deadlocks, situations where tasks mutually block each other waiting for resources, do not occur
- **Non-blocking access**: Access to resources in the AUTOSAR OS never results in a waiting state, allowing for efficient task execution without unnecessary delays.

The following code snippet illustrates the usage and locking of resources within the AUTOSAR framework, specifically highlighting the implementation of a critical section to ensure exclusive access to a particular resource:

```
#define RESOURCE_ID 1
TASK(Task1)
{
    GetResource(RESOURCE_ID);
    // Critical section, access the shared resource here...
    ReleaseResource(RESOURCE_ID);
    TerminateTask();
}
```

In this example, `Task1` acquires the resource using `GetResource()`, accesses the shared resource, then releases it using `ReleaseResource()`. During the time `Task1` holds the resource, no other task can acquire the same resource, ensuring exclusive access.

In summary, resource management in the AUTOSAR OS is a critical feature that guarantees safe and efficient sharing of resources across tasks and ISRs, paving the way for effective, concurrent execution in real-time systems.

Hooks

Hooks in AUTOSAR are callout functions that users can implement to customize and extend the behavior of the OS at specific points during its execution. These hooks serve as entry points for user-defined code and are designed to be called by the AUTOSAR OS when certain events or conditions occur. Hooks are particularly useful for detecting issues, debugging, and performing additional actions in response to critical events.

Hooks can be used in various scenarios and are typically categorized based on the specific state or event they are associated with. Some of the commonly used hooks in AUTOSAR include shutdown hooks, task execution hooks, and error hooks. Let's take a look at these:

- **Task/ISR execution hooks**: Task execution hooks are called before and after the execution of each task within the AUTOSAR OS. These hooks provide insight into the task execution flow and can be used to monitor task behavior, collect run-time information, or perform additional checks. Task execution hooks are useful for profiling, performance monitoring, and detecting anomalies or errors during task execution.

- **Shutdown hooks**: Shutdown hooks are called when the system is shutting down or entering a low power state. They allow users to perform cleanup tasks, save critical data, or gracefully shut down system components before the system powers off. Shutdown hooks can be used to trigger data storage in **non-volatile memory** (**NVM**), release resources, or ensure the system is in a safe state for shutdown.

- **Error hooks**: Error hooks are triggered when an error condition occurs within the AUTOSAR OS or when critical errors are detected. These hooks allow users to capture error information, perform error-handling routines, and take appropriate actions to mitigate the impact of errors. Error hooks can be used to log error messages, generate error reports, or initiate error recovery procedures.

Hooks are beneficial in AUTOSAR systems as they provide a means to extend the functionality of the OS and incorporate user-defined code at critical points. They enable users to monitor system behavior, gather diagnostic information, and take corrective actions when necessary. By implementing hooks, developers can enhance system observability, facilitate troubleshooting, and improve the overall reliability and robustness of the AUTOSAR-based software.

It is important to note that the implementation and usage of hooks should be carefully considered, as they introduce additional code and execution overhead. Hooks should be designed to be efficient, non-blocking, and have minimal impact on system performance. Careful testing and validation of hook functions are essential to ensure they do not introduce unintended side effects or compromise system stability.

Summary

In this chapter, we explored the role of the AUTOSAR OS in automotive software development. We discussed its architecture, RTOS, and the OSEK standard. By understanding the role of RTOS in the automotive industry, you have gained a deeper understanding of the importance of precise time and resource management. The AUTOSAR OS, with its priority-based scheduling, fast interrupt processing, and inter-task communication capabilities, equips you to develop high-performance and time-critical automotive software systems. Exploring the influence of the OSEK standard on the AUTOSAR OS has provided you with insights into addressing automotive constraints through preemptive and non-preemptive scheduling, as well as event-driven task synchronization. These features enable you to design and optimize the execution and timing requirements of tasks, ensuring optimal performance within automotive systems.

Furthermore, you have familiarized yourself with essential elements of the AUTOSAR OS, such as tasks, alarms, events, services, scheduling, resources, and interrupts. This knowledge empowers you to effectively configure and manage these components, facilitating task execution, synchronization, and resource allocation in ECUs. You can synchronize and handle state transitions of extended tasks through assigned events, enhancing the overall functionality and coordination of your software components. Additionally, understanding resource management ensures coordinated access and conflict prevention in shared resources, optimizing system efficiency and reliability.

You are now well equipped to apply your expertise in real-world automotive software development scenarios. Your understanding of the AUTOSAR OS, RTOS, task management, synchronization, and resource allocation will contribute to the design and development of robust, efficient, and safe automotive systems.

The next chapter will delve into an in-depth exploration of the application layer, the RTE, and how software components are designed within the AUTOSAR framework.

Questions

1. Describe the main objects an AUTOSAR OS consists of.
2. What is the difference between a basic task and an extended task?
3. When do you select FULL or NON as a task property in the OS configurator?
4. How can we protect a shared resource?

Get This Book's PDF Version and Exclusive Extras

Scan the QR code (or go to `packtpub.com/unlock`). Search for this book by name, confirm the edition, and then follow the steps on the page.

Note: Keep your invoice handy. Purchases made directly from Packt don't require one.

7
Exploring the Communication Stack

The **communication stack** is a central pillar of the AUTOSAR framework, intricately weaving together diverse modules to create a harmonious symphony of data exchange and functional integration. As we navigate through this chapter, our expedition will concentrate on the various modules constituting the AUTOSAR communication stack and their significance in facilitating communication protocols within an automotive **electronic control unit** (**ECU**).

At the heart of our discourse lies the intricate design, organization, and functional dynamics of these modules. Delving into their very fabric, we aspire to uncover how they interface with one another. This chapter focuses on dissecting each layer of the communication stack and shedding light on its purpose and design considerations, examining its modules, unraveling their intricate interdependencies, and introducing key terms such as messages, frame, **protocol data unit** (**PDU**), and signal.

We will be covering the following topics:

- What is the COM stack?
- COM stack overview
- Network management

Let's dive in and explore what the COM stack is and why it's one of the most essential stacks in an automotive ECU.

What is the COM stack?

Imagine you're a developer working on an ECU responsible for monitoring engine temperature in a vehicle. Your task is to ensure that the ECU can send alerts to other ECUs in the vehicle CAN network when the temperature exceeds a certain threshold.

In a traditional embedded implementation, you would need to handle the entire communication process yourself. This involves configuring the hardware interfaces (such as CAN, FlexRay, or Lin) for communication, formatting the message data, managing the timing and synchronization of data transmission, and implementing error detection and recovery mechanisms. This process requires too much low-level coding and is highly dependent on the specific hardware and communication protocols used in the system.

In AUTOSAR, communication between ECUs is abstracted through standardized interfaces and a layered communication stack. Here's how the same scenario would be handled in an AUTOSAR-compliant system:

- **Application Layer**: Your application running on the ECU generates a signal indicating the high engine temperature.

> **Note**
> A **signal** in AUTOSAR is a piece of information or data transmitted between different components or software modules within a vehicle's electronic systems. Signals can represent various types of data, such as sensor readings, actuator commands, or status information.

- **Run-Time Environment (RTE) Layer**: The signal is passed to the RTE Layer, which abstracts the application's communication needs from the underlying hardware and protocols.

- **Communication stack**: The communication stack, which consists of various layers, takes over the responsibility of packing and then transmitting the signal. It also handles the routing of the message to the appropriate destination ECU based on predefined communication routes. For example, is the message being sent over CAN, FlexRay, or any of the other automotive communication buses?

  The communication hardware interfaces, such as **controller area network** (**CAN**) controllers or Ethernet PHYs, handle the physical transmission of the message over the communication bus.

In the next section, we will examine the software architecture of the COM stack in AUTOSAR.

COM stack overview

The Communication Stack is a suite of software modules that handle communication tasks in AUTOSAR-compliant systems. It is responsible for transmitting and receiving data across various in-vehicle networks, such as CAN, **local interconnect network** (**LIN**), Ethernet, and FlexRay, among others. Its structure facilitates modular and scalable communication across different automotive ECUs.

Figure 7.1 shows the main components of the COM stack:

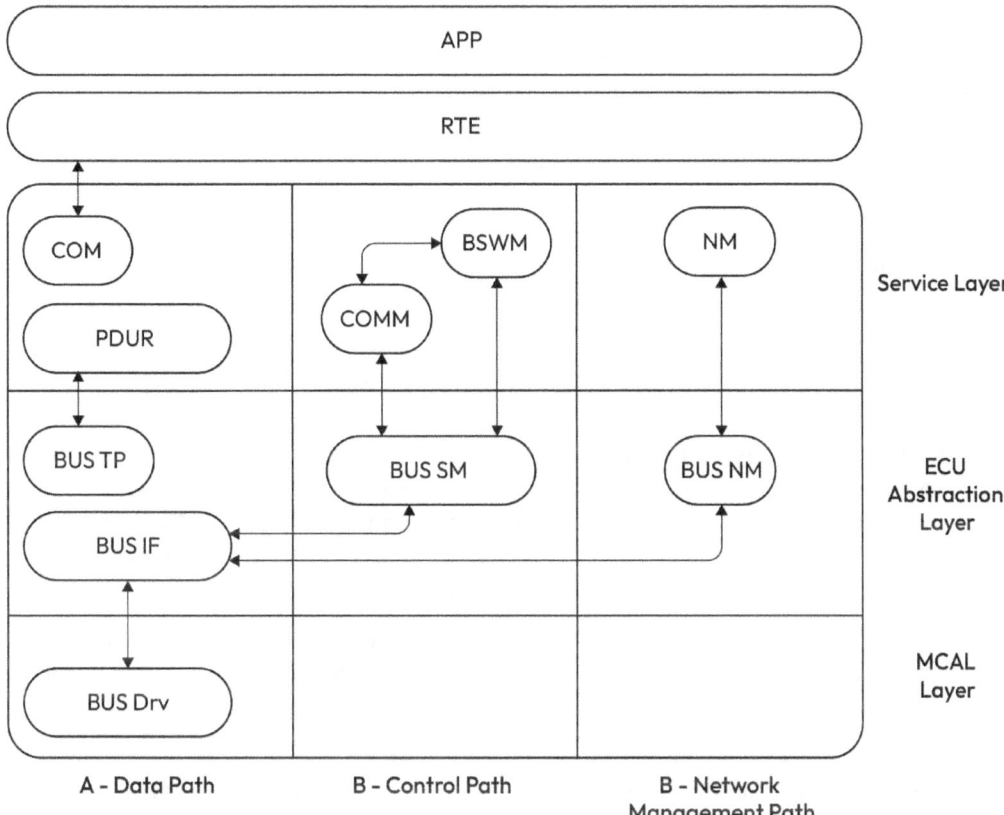

Figure 7.1 – COM layered architecture

In *Figure 7.1*, the communication stack of AUTOSAR is segmented into three main components:

- The data path handling signals and PDU and transmission (A)
- The control path handling the state of the communication channels (B)
- Network management (C)

These parts of the communication stack are distributed across various layers: the Service Layer, the ECU Abstraction Layer, and the Hardware Layer. This distribution illustrates how AUTOSAR manages communication functions, from handling raw data transmission to more complex control and **Network Management (NM)** tasks.

Here's a brief breakdown of some of the layers and their responsibilities within the AUTOSAR COM stack:

- **Application Layer**: This layer contains application-related software components. They create or consume the data that needs to be communicated. For instance, receiving the current **Vehicle Speed (VehSpd)**, or transmitting the ECU state and control signals to other ECUs.

- **RTE Layer**: As covered in *Chapter 4* and *Chapter 5*, the RTE serves as a mediator between the Application Layer and lower communication stack layers, abstracting communication details for application components.

Next, we turn our attention to the basic software, which, as mentioned earlier, is structured into three layers: Service, ECU Abstraction, and **microcontroller abstraction layer (MCAL)**. Let's now explore the modules that make up the Service Layer:

- **Communication Service Layer**

  - **COM**: It abstracts the communication details further and provides services for **protocol data unit (PDU)** handling. The COM layer's main responsibility is handling signal-based communication. COM abstracts the underlying communication protocols and hardware interfaces, allowing software components to communicate in a platform-independent manner. Later, we shall cover COM in more detail.

  - **PduR**: This module routes PDUs from one layer to another (e.g., from the COM layer to the appropriate bus-specific interface (CAN, FlexRay, LIN, etc)), or provides gateway functionality between two communication interfaces (e.g., CAN to Ethernet).

  - **BUS-TP**: The Transport Protocol Module in the COM stack manages reliable data transmission between ECUs, handling segmentation, flow control, error detection, and addressing (e.g., CAN-TP).

  - **BUS State Manager (BUS-SM)**: The BUS-SM is a protocol-specific module in AUTOSAR's COM stack. It manages communication mode transitions for each network, based on requests from the AUTOSAR **Communication Manager (ComM)**. It handles mode change requests and state change notifications, manages bus-specific state machines, and performs error handling (e.g., bus-off conditions). It also notifies other modules such as **Basic Software Module (BswM)** and ComM about state changes, ensuring synchronized communication across the vehicle network (e.g. **CAN State Manager (CanSM)**, **Ethernet State Manager (EthSM)**).

Next, we will explore the ECU Abstraction Layer contents of the stack:

- **Communication ECU Abstraction Layer**:

  - **BUS Interface (BUS IF)**: These are network-specific modules for various in-vehicle communication protocols, such as CAN, LIN, FlexRay, and so on. It serves as the interface between the higher layers of the communication stack and the hardware-specific drivers. BUS-IF handles the adaptation of protocol-specific communication services for different network types. It translates the generic communication requests from the **PDU Router (PduR)** and **Communication Manager (ComM)** into protocol-specific commands for the respective bus drivers. Examples include **CAN Interface (CanIf)**, **LIN Interface (LinIf)**, and **FlexRay Interface (FrIf)**.

Then lastly, the MCAL Layer part for the stack:

- **Communication Driver Layer**:
  - **Bus drivers**: These are the lowest level in the communication stack, directly dealing with the hardware interfaces of the communication channels. Examples include the CAN Driver, LIN driver, and FlexRay driver.

> **Note**
> Network management will be discussed in more detail later, but these modules are key components, COMM, BUS-SM, and CanSM, which provide state machines for managing communication channels and overseeing initialization, configuration, and control of communication activities. They handle data transmission, monitor communication status, manage errors, control controller states, and manage sleep-wake-up operations.

Together, these layers and modules ensure that data can be effectively packaged, transmitted, received, and unpacked across various ECUs in an automotive system, allowing for efficient and reliable communication in the vehicle.

By now, we have a basic understanding of the functions of the communication stack. Before exploring the stack deeper, let's clarify some of the fundamental concepts and terms.

An overview of basic COM concepts

To truly grasp the nuances and intricacies of the COM modules, one must first familiarize oneself with its foundational concepts. This section explains key concepts, setting the stage for a thorough understanding of AUTOSAR COM. We will explore the following:

- **Signals**: The basic data units in AUTOSAR, which can represent sensor readings, actuator commands, or other pieces of information. They become the atomic units of communication within and between software components.

> **Note**
> **VehSpd**, potentially represented as a `uint16` data type, and `OdometerCount` might be represented as `uint32` are examples of signals. A signal can be any type of data that is incorporated into a message. For instance, in a CAN message, the initial two bytes might denote VehSpd, while subsequent bytes could signify another signal, perhaps referred to as signal *X*.

- **Signal groups**: This refers to a collection of signals that are logically related and are treated as a single entity for certain operations, particularly during communication processes. Signal groups are useful for ensuring data consistency when multiple signals need to be transmitted together, enabling more organized and efficient data exchange.
- **Service Data Unit (SDU)**: This is the payload or data content that is passed between layers in a communication stack. Each layer in the AUTOSAR software stack may add its own headers to the SDU, transforming it into a PDU for the next layer.
- **PDU**: An encapsulated data packet transmitted between communication layers. It comprises both payload (data) and **Protocol Control Information (PCI)**.
- **PCI**: It refers to headers and metadata for managing the data packet flow through the AUTOSAR stack layers and specifies the SDU's next destination. *Figure 7.2* provides an illustration of the PDU structure and its decomposition through the COM stack.

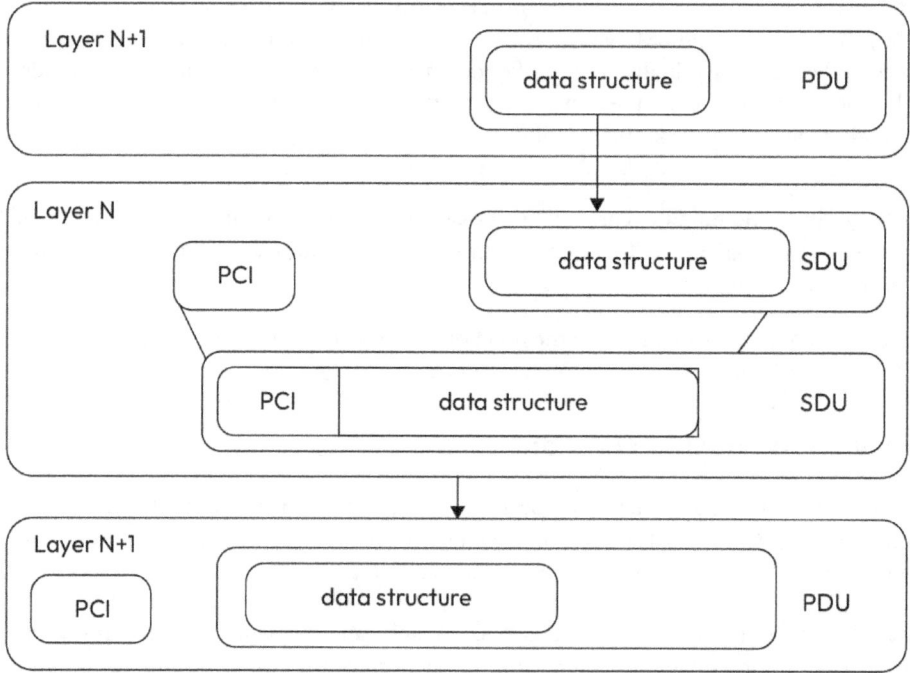

Figure 7.2 – PDUs and PCIs

To better understand the PDU, SDU, and PCI, let's consider the following example for data communication from PduR through the **CAN Transport Protocol (CanTp)**, to CanIf, and to CAN Driver, where a message assumes its content is ABCDEFGHIJ.........., this is sent, it starts as a PDU at one layer, becomes the SDU for the next, and gets segmented and appended with PCI, such as sequence numbers within the CanTp scope.

Here is the flow:

1. A user (COM) provides the PduR with an SDU (request to send a PDU).

    The PduR receives an SDU with content, such as ABCDEFGHIJ.......... Then, it packs the SDU along with PCI (routing info in this case), and the resulting PDU is passed to a lower layer, in our case, CanTp, and this PDU would be considered by the CanTp as an SDU.

    SDU: This is the raw data or message content that the PduR got from an upper layer (e.g., COM says 1 k bytes of data). ABCDEFGHIJ.......... The PCI at the PduR level is empty, nothing is added to the SDU, even without explicit PCI, and routing the SDU to a lower-layer module (such as CanTp) acts as implicit PCI.

    PCI: Empty (not needed at this stage).

    Resulting PDU content: ABCDEFGHIJ..........

2. CanTp treats the received PDU as its SDU, which is essentially the payload it needs to process.

    CanTp adds PCI to the SDU. This PCI might include a Frame type, sequence number, addressing information, and data size to manage the data's segmentation and reassembly at destination reception; we will refer to it as Other in this case.

    SDU: ABCDEFGHIJ..........

    PCI added: Type Seq Other: This refers to the Frame type, sequence number, and other metadata that could be needed for a PCI.

    Note that the following data is segmented:

    - **Segment 1**: Type Seq Other ABCDEF
    - **Segment 2**: Type Seq Other FGHIJ....
    - **Segment N**: Type Seq Other ....

3. CanTp provides the segment data each as an individual PDU.

    CanIf receives these segments from CanTp. It adds a CAN ID, the CAN message identifier, to each segment:

    I. **SDU**: Type Seq Other ABCDEF.

    II. **PCI added**: CANID.

    III. **Segment 1 PDU**: CANID Type Seq Other ABCDEF

    IV. **PCI added**: CANID

    V. **Segment 2 PDU**: CANID Type Seq Other   FGHIJ....

    Then, the CAN Driver obtains CanIf PDUs as its own SDU and transmits it over the physical bus.

In summary, the original message (ABCDEFGHIJ..........) undergoes segmentation as it's very large and can't fit within a single CAN frame, and various protocol layers add specific control information to ensure the message is transmitted correctly over the CAN network. Each layer adds its own PCI if needed before passing the data to the next layer or the CAN network for transmission.

Figure 7.3 provides a clearer sequence diagram of how the previously mentioned process goes in each layer:

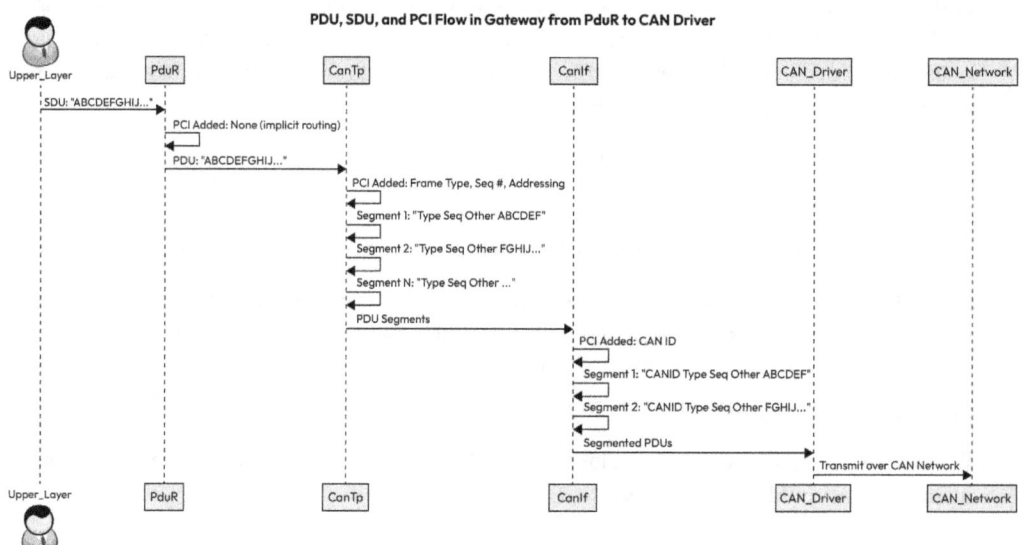

Figure 7.3 – UML sequence showing data flow from PduR to CAN Driver, with each layer adding PCI and segmenting data before final CAN network transmission

Gatewaying is the mechanism by which signals or PDUs can be routed or mapped from one communication channel to another. For example, a message from channel CAN A will be sent/redirected with the same content to another communication bus. *Figure 7.4* shows a sequence diagram example where a frame is received in CAN and then gatewayed to Eth and FlexRay.

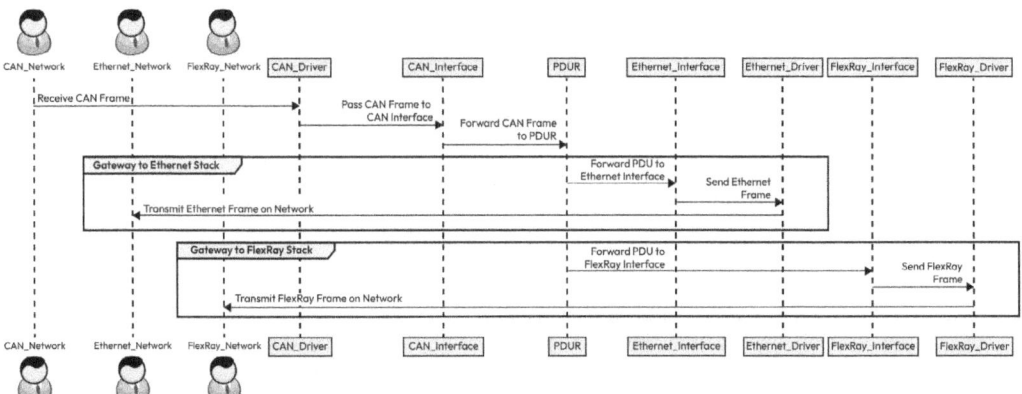

Figure 7.4 – Gateway of a PDU from the CAN network to ETH and FlexRay networks

We are now prepared to take a deeper dive into the core modules of the COM stack in the case of the CAN network. I will not explore the details of other communication buses, but the same concepts can generally be applied when working with them. Let's begin our exploration of the COM module.

COM module

The COM module in AUTOSAR serves as an intermediary layer between the Application Layer and the underlying communication hardware. Its primary function is to manage the sending and receiving of data (in the form of signals) between software components in the system and the communication infrastructure.

The features offered by the COM module include the following:

- **Signal handling**: The COM module manages the abstraction of application data into signals. Signals can represent everything from sensor readings to control command values.

- **Signal grouping**: Multiple signals can be grouped together into what's known as **signal groups**. This allows for the efficient bundling of related signals, ensuring they're transmitted together and preserving data consistency.

- **PDU handling**: The COM module takes signals (or signal groups) and packages them into PDUs for transmission. Similarly, it extracts signals from received PDUs.

- **Notification mechanisms**: The COM module can notify the Application Layer about the successful transmission or reception of signals, or the occurrence of errors.

Figure 7.5 shows how the AUTOSAR COM module packages signals from various SWCs into a PDU and sends them to the lower communication layers for transmission across the network.

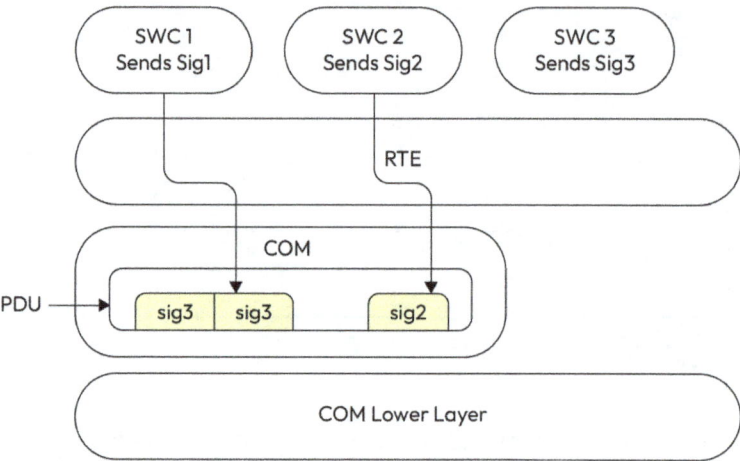

Figure 7.5 – COM module packing signals into a PDU

Now, let's jump into the details of sending and receiving operations within the COM module, focusing on how data is managed through the process of packaging into PDUs and how the COM module handles these operations.

Sending and reception in COM

The COM module manages the data flow between ECUs by packaging data into PDUs and handling the transmission and reception according to defined transmission modes (Direct, Periodic, Mixed, If Active, or None):

- **Sending**: When the Application Layer has data to send, it passes this data to the COM module, which abstracts it as a signal. The COM module then maps this signal into a specific area within the PDU, which is sent to the lower layers for transmission on the physical bus.

- **Reception**: For incoming data, the COM module extracts signals from the received PDU and provides the actual data to the Application Layer. During this process, it might use filtering mechanisms to decide which signals should be processed.

Now that we understand how the COM module manages the sending and reception of data by packaging and unpacking PDUs, it's time to explore the different transmission modes that dictate how and when this data is transmitted. Understanding these modes is helpful in optimizing communication efficiency and ensuring reliable data flow within automotive systems. Let's dive into the various transmission modes provided by the AUTOSAR COM module.

Transmission modes in COM

These modes dictate how and when data is transmitted, catering to different requirements of timing, event triggering, and data aggregation. Understanding these transmission modes is helpful for optimizing communication efficiency, ensuring data consistency, and managing network bandwidth in automotive systems. The AUTOSAR COM module provides distinct transmission modes for PDUs:

- **DIRECT (N-Times):**

    Description: In this mode, the data (PDU) is transmitted immediately upon update if the triggering condition is met. For instance, if the DIRECT (N-Times) count is set to 3, the COM module will transmit three PDUs upon receiving signals from RTE.

    Importance: This is useful for critical data that must be sent without delay, such as emergency vehicle signals.

    Figure 7.6 shows when a PDU is marked as updated, transmission happens right away.

    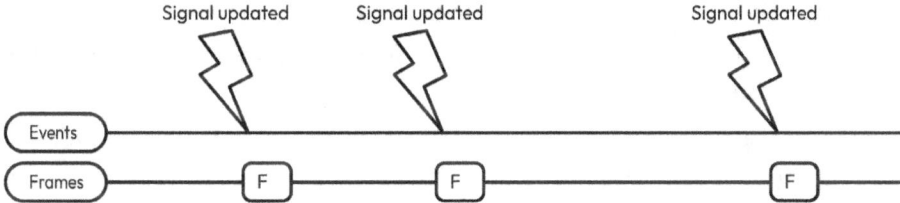

    Figure 7.6 – PDU transmission mode is Direct with N = 0

    Figure 7.7 shows the same setup when N is set to 3, thus, this frame is visible three times in the communication bus.

    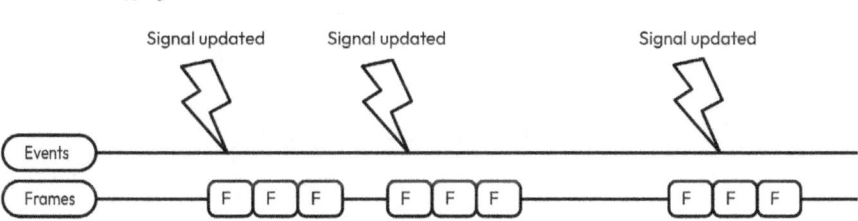

    Figure 7.7 – PDU transmission mode is Direct with N = 3

- **Periodic:**

    Description: Data is sent at predefined time intervals, regardless of changes in the data or other triggering conditions.

    Importance: It ensures regular updates, which is vital for systems that rely on steady and predictable data inputs.

Figure 7.8 shows that in Periodic mode, a PDU is always requested for transmission at time *t* with the most recent data, regardless of whether the PDU has been updated.

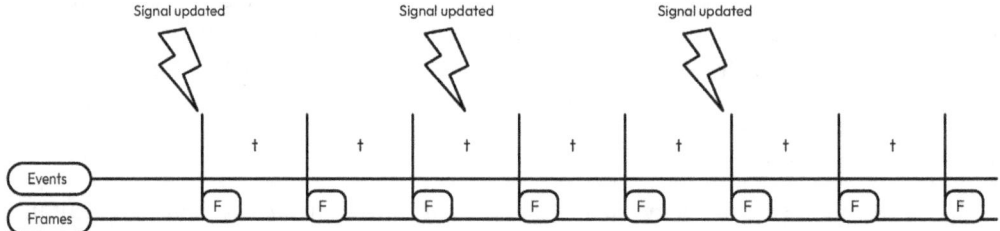

Figure 7.8 – PDU transmission mode is Periodic Time = t

- **Mixed**:

    Description: A combination of Direct and Periodic transmission. Data is transmitted immediately when it changes, but also periodically if there are no changes.

    Importance: It balances the need for immediate transmission upon change with the reliability of periodic updates.

    Figure 7.9 illustrates that in mixed mode, a PDU is always transmitted within its period of time and also when it's triggered (updated).

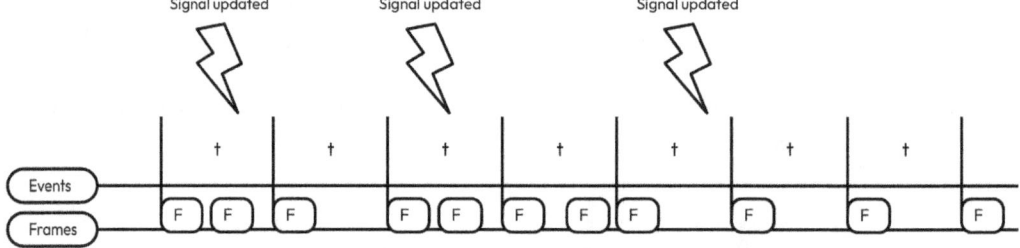

Figure 7.9 – PDU transmission mode is Mixed Time = t

- **None**:

    Description: No automatic transmission; data must be requested. This may be by the communication bus through Trigger Transmit methods.

    Importance: Useful for reducing network traffic by eliminating unnecessary data transmissions.

Each of these modes serves specific purposes and choosing the appropriate transmission mode is critical for ensuring the performance and reliability of automotive communication systems. It impacts how effectively a vehicle's electronic systems communicate, which in turn affects overall vehicle operation, safety, and efficiency.

Transfer properties for signals and signal groups

The transfer properties of signals define how and when a signal should be transmitted within the communication stack. Two commonly used transfer properties are PENDING and TRIGGERED. Let's understand their roles:

- PENDING: When a signal is set with the PENDING transfer property, it indicates that the signal is ready to be transmitted but awaits certain conditions or events before the actual transmission can occur. The signal, in essence, remains in a standby or "waiting" state until those conditions are met.

  This property is often useful in scenarios where the system might need to batch-process or aggregate several signals before sending them out. Alternatively, it can be used when the signal needs to be transmitted after certain events, such as a response from another module or after a timeout.

- TRIGGERED: A signal with the TRIGGERED transfer property is immediately scheduled for transmission once it's updated or when a specific triggering event occurs. There's no holding back or waiting – as soon as the conditions defined by the TRIGGERED property are met, the signal is sent to the lower layers for transmission.

  This property ensures real-time or near-real-time communication and is especially valuable in systems where timely data transmission is needed, such as safety-critical applications.

Let's explore specific examples to illustrate how transfer properties of signals in AUTOSAR are associated with different transmission modes, demonstrating their practical application in automotive systems.

Example 1 – Triggered transfer with Direct Transmission Mode

Scenario: An airbag deployment signal in a vehicle:

- **Transfer property**: TRIGGERED
- **Transmission mode**: Direct
- **Description**: The airbag deployment signal needs to be communicated instantly to ensure the safety systems react without delay in the event of a collision. The TRIGGERED transfer property is used because the signal must be sent the moment its value changes (from not deployed to deployed). Using the Direct Transmission Mode, this signal is transmitted immediately to ensure a rapid response.

Example 2 – Pending transfer with Periodic Transmission Mode

Scenario: Engine temperature monitoring:

- **Transfer property**: `PENDING`
- **Transmission mode**: Periodic
- **Description**: The engine temperature is a vital parameter that needs consistent monitoring but does not typically require immediate action unless there are critical changes. The `PENDING` transfer property fits as the data is regularly updated and sent at fixed intervals (e.g., every 30 seconds), ensuring ongoing monitoring without the need for immediate transmission upon every minor change.

Having explored the intricacies of transmission modes and their impact on signal management, let's now turn our attention to another vital feature of the COM module in AUTOSAR: signal filtering. This functionality significantly enhances data handling and boosts system performance by selectively processing communication data.

Signal filtering in the AUTOSAR COM module

The AUTOSAR COM module can apply specific filtering criteria to signals received, thereby enabling certain signals to be selectively processed. While the COM module employs filtering mechanisms for transmission mode conditions, it's noteworthy that it doesn't exclude signals during transmission.

The various filtering algorithms within the COM module, labeled as `ComFilterAlgorithms`, include the following:

- `ALWAYS`: The module doesn't apply any filtering, allowing every message to proceed
- `NEVER`: This filter consistently prevents messages from being processed
- `MASKED_NEW_EQUALS_X`: Only messages with a specific masked value matching a designated value are allowed
- `MASKED_NEW_DIFFERS_X`: Messages that don't match a designated masked value are allowed
- `MASKED_NEW_DIFFERS_MASKED_OLD`: Only messages displaying a change in the masked value when compared to the previous value are processed
- `NEW_IS_WITHIN`: Messages are processed if their value lies within a specified range
- `NEW_IS_OUTSIDE`: Only messages with values that fall outside a specified range are permitted
- `ONE_EVERY_N`: This filter allows one message to pass for every set of *N* messages received

Through these filtering mechanisms, the COM module offers enhanced flexibility and selectivity in managing incoming signal data.

Having understood the transmission modes and properties in COM, it's time to explore another key aspect of the COM module that influences critical data processing and can impact the overall system's performance.

COM processing modes

One of the key aspects of COM processing involves deciding whether to handle communication requests/notifications PDUs immediately or defer them for later processing in the next processing slot of the COM module. This choice significantly impacts the efficiency, responsiveness, and resource utilization of the vehicle's electronic control systems.

Immediate processing

Immediate processing in the COM module refers to the immediate handling of data once it is received or ready to be sent. This method is typically employed when the data's timeliness is critical to the system's operation.

Use case: A prime example of immediate processing is the transmission of safety-critical signals, such as airbag deployment notifications. When an airbag sensor detects a collision, the signal indicating deployment must be transmitted instantly to ensure that the airbag deploys without delay.

Deferred processing

Deferred processing involves storing data temporarily and processing it at the next available processing slot (the next main function invocation). This approach is used when immediate handling is not necessary and can be scheduled to optimize system resources.

Use case: An example of deferred processing is the periodic transmission of non-critical telemetry data, such as fuel consumption or regular diagnostic information. In this case, the COM module can collect data over a period and process a queue of data at a later point, benefiting the usage and availability of CPU time.

To complete our understanding of the COM module, it's beneficial to explore the commonly used APIs. As a developer, knowing these APIs will not only enhance your ability to interact with the COM module effectively but also simplify the debugging process, giving you a deeper insight into the module's operations.

COM modules APIs

In this section, we will explore various APIs provided by the COM module, focusing on how they are typically utilized by the RTE to receive and update signals within AUTOSAR systems:

> **Note**
> Please note that the APIs listed here represent only a selected subset of the CAN APIs. For comprehensive details and specifications, always refer to the official AUTOSAR standard documentation.

- `Com_RxIndication`:

    Purpose: This API is called by the lower layers (such as CAN and LIN drivers) when a PDU has been received. It notifies the RTE that data is available.

    Parameters:

    - `PduId`: This is the identifier of the received PDU
    - `PduInfoPtr`: This points to the structure containing the received data and length information

- `Com_TxConfirmation`:

    Purpose: Used to confirm that a PDU has been successfully transmitted. This API is invoked after the successful transmission of a PDU.

    Parameters:

    - `PduId`: This is the identifier of the PDU that was transmitted

- `Com_ReceiveSignal`:

    Purpose: This allows the application to read the latest data received for a specific signal.

    Parameters:

    - `SignalId`: This is the identifier of the signal whose data is being requested
    - `SignalDataPtr`: This points to where the signal data should be stored

- `Com_SendSignal`:

    Purpose: This allows the application to update the data of a signal within the COM module. When the Application Layer needs to transmit new data over the communication network, this function enables the data to be sent through the COM module.

    Parameters:

    - `SignalId`: This is the identifier of the signal whose data is being updated
    - `SignalDataPtr`: This points to where the signal data should be stored

COM stack overview

Having explored the COM module, let's now shift our focus to the PduR module, another key component in the COM stack, which plays the main role in mediating between different communication modules and the COM layer.

PDU-Router (PduR) Module

PduR stands as a central module within the AUTOSAR communication stack. It facilitates the routing of PDUs between layers and modules, acting as a mediator to ensure data reaches its intended destination. In essence, PduR allows for the decoupling of communication layers, meaning that upper layers, such as the COM module, don't need to be explicitly aware of the specifics of the lower-layer network implementations, such as CAN, LIN, or FlexRay.

Let's explore the main concepts for PduR.

Key features within the PduR module

The PduR module in AUTOSAR offers several functionalities that support efficient data communication across various network protocols. The following are the key features that define the PduR module's capabilities:

- **Routing**: At its core, PduR is responsible for routing PDUs from a source to a destination module. This encompasses activities such as taking PDUs from the COM layer and directing them to the appropriate bus-specific Interface Layer (such as CanIf for the CAN network).

Figure 7.10 demonstrates the routing functionality of the PduR module, showing how a PDU from the COM module is effectively routed to the CAN bus.

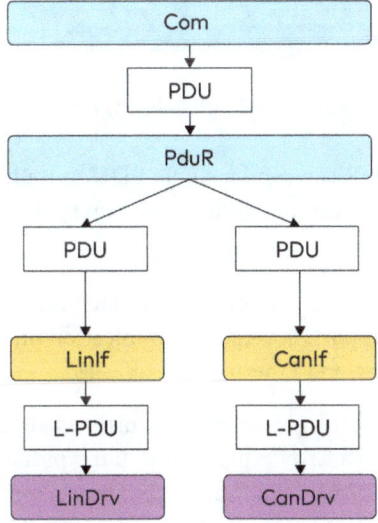

Figure 7.10 – PduR routing flow

- **Static configuration**: PduR relies on a static configuration, meaning that all routes (i.e., the paths data takes from one module to another) are determined at compile time. This ensures that there's no dynamic decision-making overhead during runtime, which benefits performance and predictability.
- **PDU fan-out**: It supports the distribution of a single incoming PDU to multiple destinations. This is particularly useful when the same PDU needs to be processed by multiple upper-layer modules.
- **Gatewaying**: In scenarios with multiple communication networks (e.g., a vehicle with both CAN and FlexRay), PduR can act as a gateway, routing PDUs from one network protocol to another.

Figure 7.11 illustrates the gateway functionality of the PduR module, depicting how a message originating from the LIN bus is routed through the PduR level to the CAN bus.

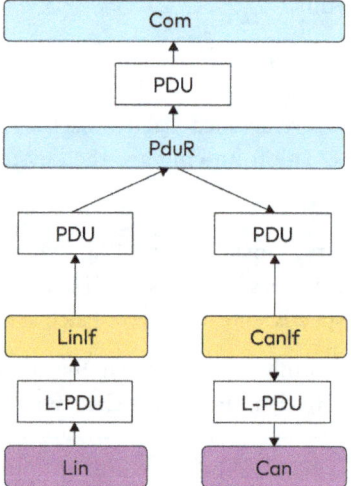

Figure 7.11 – PduR gateway flow

- **Multicast capabilities**: PduR can send a single PDU to multiple receiving modules. This feature is particularly useful in systems where the same data needs to be distributed across various components.

Up to this point, the configuration of the communication stack varies based on the communication protocol employed. AUTOSAR supports several communication protocols relevant to automotive systems, such as CAN, LIN, FlexRay, and Ethernet.

Given that each protocol requires distinct handling, the corresponding hardware Abstraction Layer modules also differ. For instance, in a CAN setup, you would typically find modules such as CanIf, **CAN Network Management (CanNm)**, CAN, and **CAN Transceiver Driver (CanTrcv)**. Similarly, for FlexRay, the relevant modules include **Flexray Interface (FrIf)**, **Flexray Network Management (FrNm)**, and **Flexray Driver (Fr)**, while Ethernet-related functionalities are supported by modules such as **Service Discovery (SD)**, **Scalable Service-Oriented MiddlewarE over IP (SomeIp)**, **Transmission Control Protocol/Internet Protocol (TcpIp)**, and **Socket Adapter (SoAd)**.

For clarity, we will focus on CAN-related modules to better understand the functionality of the COM module.

CAN modules

The AUTOSAR CAN stack is designed as a multi-layered architecture that abstracts the hardware details and provides standardized interfaces between the layers. At a high level, it consists of CanIf, the CAN Driver, CanTrcv, CanNm, CanSM, and the **CAN Transport Protocol (CanTP)**.

Let's explore each of these modules individually, examining their features and use cases.

CanIf module

The CanIf module in AUTOSAR abstracts the complexities of the CAN hardware from the upper layers by providing a standardized interface for communication, allowing application developers to interact with the network without needing to manage or understand the specific details of the CAN hardware, such as baud rate or scheduling logistics.

> **Note**
> The upper layer could be the PduR module, the CanTp module, or another complex device driver module that needs to transmit a PDU over the CAN network.

The following are some of the CanIf features:

- **Role**: It acts as an intermediary between the higher layers, such as the PduR and the lower layers (the CAN Driver).
- **Multiple drivers**: CanIf is designed to handle multiple CAN controllers/drivers. When you have multiple CAN controllers in each ECU, CanIf consolidates and manages access to these controllers.
- **Message buffering**: It handles the buffering of messages using hardware or software-based message boxes, ensuring efficient use of available resources.
- **Filtering**: CanIf performs message filtering based on identifiers. It ensures that the upper layers receive only those messages that are relevant to them.

The next section introduces some of the important CanIf APIs.

CanIf APIs

Please note that the APIs listed here represent only a selected subset of the CanIf APIs. For comprehensive details and specifications, always refer to the official AUTOSAR standard documentation.

- `CanIf_Transmit`:

    Purpose: It requests the transmission of a CAN PDU.

    Parameters:

    - `CanTxPduId`: This is the identifier of the CAN PDU that should be transmitted
    - `PduInfoPtr`: This points to the PDU structure containing the data and length of the CAN PDU

    Description: This function is used to transmit data over the CAN network. It manages the mapping from the PDU ID to the corresponding CAN hardware object and handles the queuing and prioritization of messages.

- `CanIf_RxIndication`:

    Purpose: It indicates the reception of a CAN PDU.

    Parameters:

    - `Hrh`: The hardware receive handle, identifying the CAN hardware receiving the message
    - `CanId`: This is the identifier of the received CAN PDU
    - `CanDlc`: The data length code of the received CAN PDU
    - `CanSduPtr`: This points to the received data

    Description: This API is called by the CAN Driver to notify CanIf about the reception of messages. CanIf then processes this information and passes it up to the respective upper layers.

- `CanIf_TxConfirmation`:

    Purpose: It is a confirmation callback for the successful transmission of a CAN PDU.

    Parameters:

    - `CanTxPduId`: This is the identifier of the CAN PDU that has been successfully transmitted

    Description: This callback is used by the CAN Driver to inform CanIf that a PDU has been successfully sent, allowing CanIf to perform any post-transmission handling, such as notifying upper layers or clearing buffers.

In summary, CanIf in AUTOSAR standardizes and simplifies the handling of CAN network communications for automotive applications, making it easier to develop and integrate various vehicle functions.

Lastly, in the CAN stack, we will explore the CAN Driver.

CAN Driver module

The **CAN Driver** is a critical module responsible for managing the lower-level operations of the CAN hardware. This module acts as the interface between the hardware-specific communication needs of the CAN controller and the higher-level software functions, allowing applications and other software modules to communicate over the CAN network without needing to manage the detailed hardware complexities and handle the actual message transmission and reception.

One important aspect of the CAN Driver is its processing modes, which dictate how data is handled and transmitted over the network. Here's an overview of the processing modes for the CAN Driver in AUTOSAR:

- **Interrupt mode**: In this mode, the CAN controller generates an interrupt for events, such as message reception or transmission completion. This mode is preferred when real-time processing is required.
- **Polling mode**: The CAN Driver continuously checks or *polls* the status of the CAN controller to identify any events. This mode can be simpler in terms of implementation and is typically used in less time-critical applications.

Choosing between these modes depends on factors such as system load, real-time requirements, and application complexity. While interrupt mode offers a timely response, polling can be more deterministic in systems with predictable loads.

Each CAN message has a unique identifier, which determines its priority on the bus. The CAN Driver can be configured to filter incoming messages based on their IDs, thus reducing unnecessary processing for irrelevant messages.

Then, mailboxes come into play, which are buffers in the CAN controller, used for storing incoming and outgoing CAN messages. Depending on the hardware, there could be separate mailboxes for transmission and reception, or they could be combined and used for either purpose. They ensure that the CAN Driver can handle multiple messages concurrently, improving system efficiency.

Similarly, as for other modules, I would like to present some of the key APIs within the CAN Driver.

CAN Driver APIs

Please note that the APIs listed here represent only a selected subset of the CAN APIs. For comprehensive details and specifications, always refer to the official AUTOSAR standard documentation; let's explore the most common ones:

- `Can_Write`:

    Purpose: This initiates the transmission of a CAN PDU.

    Parameters:

    - `Hth`: This is the hardware transmit handler, which denotes the message buffer in the CAN hardware
    - `PduInfo`: This points to the PDU structure containing the ID, length, and data pointer of the CAN message

- `Can_Receive`:

    Purpose: This handles the reception of CAN messages. This API is often implicitly called by the CAN **Interrupt Service Routine (ISR)** and is not typically used directly by the application.

    Parameters:

    - `Hrh`: This is the hardware receive handler, representing the message buffer in the CAN hardware that has received a message
    - `CanId`: This is the identifier of the received CAN message
    - `CanDlc`: This is the data length code of the received message
    - `CanSduPtr`: This points to the received data

Having thoroughly explored the components of the communication stack, let's now recap with a practical example.

Exploring an example of sending a CAN message

To grasp the functionality of the COM stack within AUTOSAR, let's examine a practical scenario: an application transmitting a message, say, a directive to modify the angle of a car's side mirrors. *Figure 7.12* provides an illustration of the data flow from a SWC to the CAN Driver.

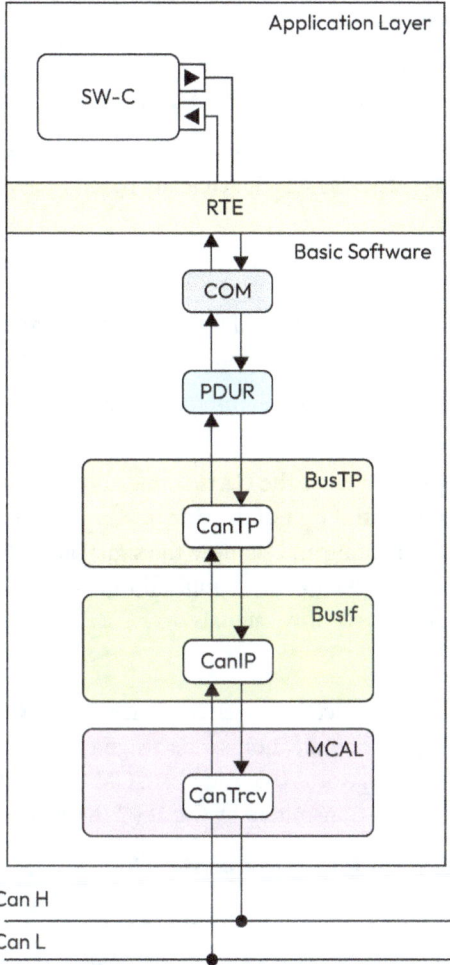

Figure 7.12 – Signals flow in COM

Let's understand what is going on:

1. **Application Layer**: This is the starting point of the process. The user interacts with the car's infotainment system and chooses to adjust the angle of the side mirrors. The Application Layer contains the software components that implement specific functionalities of the car, in this case, the side mirror adjustments.
2. **RTE**: RTE acts as an intermediary between the Application Layer and the COM module. It abstracts the communication details, enabling software components to communicate as if they were in a single address space. In this case, it takes the angle command and prepares it for transmission by passing it to the COM layer.

3. **COM**: This received the signal (the desired angle value) from the RTE layer:

    I. **Signal handling**: The COM layer will take this signal and map it to a PDU. If multiple signals need to be sent together, they can be packed into a single PDU.

    II. **Transmission mode**: The COM layer will decide when to send this PDU. This can be immediate (`SEND_IMMEDIATE`), cyclically based on a timer, or upon a change in signal value.

4. **PduR**: The data from the COM layer includes the packaged PDU.

5. **Routing**: PduR takes the PDU from the COM layer and routes it to the appropriate bus interface based on its configuration. The decision is made during compilation time and is static.

6. **CanIf**: This layer abstracts specifics of the bus protocol, translating general PDUs into bus-specific data frames.

7. **CAN Driver**: The CAN Driver takes the CAN frame and sends it as electrical signals over the CAN bus to the motor control unit of the side mirror. The Bus Driver interfaces directly with the physical layer, modulating and sending the signal across the actual communication medium (wires, optical fibers, etc.). In our example, if we're using CAN, the CAN Driver takes the CAN frame and sends it as electrical signals over the CAN bus to the motor control unit of the side mirror.

Figure 7.13 illustrates the CAN communication stack sequence in AUTOSAR, providing the flow of transmission of the CAN message the application sends a signal via `Rte_Send_Signal_a()`, which is processed through RTE (`Com_Send(Signal)`), COM (`PduR_Transmit()`), and PduR (`CanIf_Transmit()`), and finally transmitted by CanIf (`Can_Write()`).

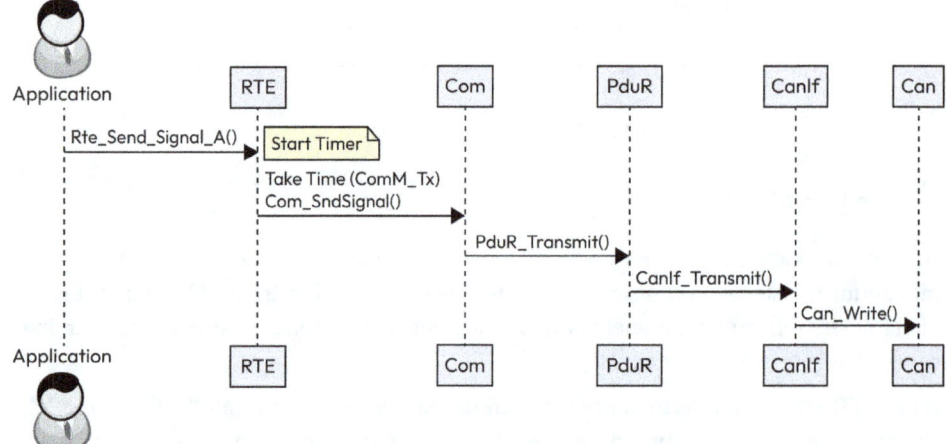

Figure 7.13 – Sequence diagram transmitting a signal

Now that we have explored the COM stack data transmission and reception path, let's try to unravel the control and Nm path.

Network management

Network management in AUTOSAR includes functions aimed at controlling and coordinating communication between ECUs across different network protocols. It handles tasks such as network configuration, fault detection and handling, and ensuring the overall integrity and reliability of data exchange. By overseeing these aspects, Nm helps maintain optimal network performance and reduces power consumption by managing the active and sleep states of ECUs. This enhances overall system performance and responsiveness.

In the next section, we shall explore the key modules to handle network management.

Key modules in network management

ComM acts as a higher-level manager that coordinates with other Nm modules, such as the **Network Management (Nm)** and various bus-specific network management modules (e.g., CanNm for CAN, LinNm for LIN), to ensure efficient and reliable communication system operations.

ComM manages various states of the communication channels:

- **Full communication mode**: In this state, the network allows active communication where ECUs can send and receive messages. ComM ensures that the network remains in this mode as long as there are active communication requests from the ECUs.

- **Silent communication mode**: Used primarily in bus systems such as CAN, where the ECUs need to listen to the bus but do not send any messages.

- **No communication mode**: In this state, the network is effectively shut down to minimize power consumption, used especially when the vehicle is parked. The network can be woken up by specific wake-up events.

> **Communication channel**
>
> Communication channels are defined within the AUTOSAR communication stack, which encompasses various communication protocols and services, such as CAN, LIN, FlexRay, Ethernet, and others, each suited for different types of data transmission needs within the vehicle.

It also collaborates with Nm modules to manage network state transitions based on ECU communication needs. It processes Application Layer requests for communication mode changes, assessing these against predefined rules and managing them with timers for stability. Additionally, ComM handles wake-up signals to transition the network from low power to active states, ensuring alignment with the vehicle's operational conditions.

Next, we can briefly discuss the CanNm as a core pillar of Nm.

The **CanNm** module is responsible for managing the state transitions and overall operation of the CAN network within the automotive system.

CanNm manages various states of the CAN network, including the following:

- **Bus-Sleep**: In this state, the CAN network is inactive, conserving power to minimize energy consumption when not in use.
- **Ready Sleep**: This state indicates that the CAN network is ready to enter the Bus-Sleep state but is currently active.
- **Normal Operation**: The CAN network is fully operational and actively transmitting and receiving messages between ECUs.
- **Bus-Off**: This state occurs when a node detects an error on the CAN bus, leading to temporary disconnection from the network. CanNm handles the recovery process to restore Normal Operation.
- **Prepare Bus-Sleep**: CanNm prepares the CAN network to transition to the Bus-Sleep state by completing necessary tasks before entering the low-power mode.

Figure 7.14 illustrates the state transitions for the previously mentioned states.

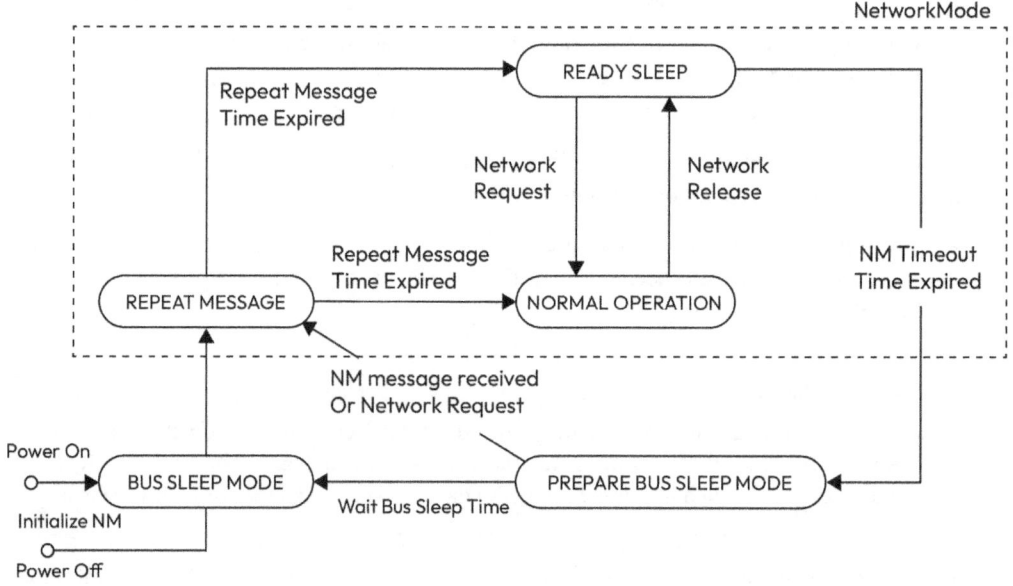

Figure 7.14 – CanNm state machine

When an ECU wakes up, it typically enters a state where it waits for Nm messages from other ECUs to synchronize its operation with the rest of the network. This state is known as the Repeat Message State. During this state, the ECU listens for Nm messages and may set a timer to wait for a certain duration before proceeding.

For example, let's consider a scenario where an ECU wakes up and enters the Repeat Message State. It waits for Nm messages from other ECUs to ensure that it synchronizes its activity with the rest of the network. If it receives Nm messages within a specified timeframe, indicating that other ECUs are active and ready to communicate, it transitions to the Normal Operation state.

However, if the ECU stops receiving Nm messages from other ECUs, it interprets this as a signal that no other ECUs want it to be active, possibly indicating a synchronized sleep scenario. In such cases, the ECU transitions through intermediate states, such as Bus-Ready Sleep, and eventually enters the Bus-Sleep state to minimize power consumption.

Synchronized sleep ensures efficient power management and network operation. By synchronizing the sleep periods of all ECUs within the network, power consumption can be minimized collectively, leading to improved energy efficiency of the entire system. Additionally, synchronized sleep prevents unintended wake-up events, where an active ECU may inadvertently wake up other ECUs unnecessarily, disrupting the network's operation and potentially wasting energy.

> **Note**
> So far, we have been describing CanNm and Nm within the context of a CAN network. It's important to note that similar concepts also apply to other communication protocols, such as FlexRay and Ethernet. While there are minor differences in implementation details and specific features, the core principles of Nm remain the same across these protocols.

In summary, the process from ECU wake-up to entering synchronized sleep involves listening for Nm messages, transitioning through intermediate states based on network activity, and ultimately entering sleep mode collectively with other ECUs to optimize power consumption and network operation.

Summary

In our exploration of the AUTOSAR COM chapter, we have dived deeply into the fundamental components and mechanisms to build reliable communication within the AUTOSAR framework. We began by unraveling the intricacies of the COM module, elucidating its paramount role in facilitating data exchange across the system. Our journey also introduced us to the significance of signals, these atomic elements of data communication, and how they play into the larger picture of data transfer. One cannot discuss COM without reference to the PduR module. We shed light on its multifaceted features, emphasizing its role as a mediator in routing and transforming data between software modules. Providing a practical perspective, the CAN stack example served as an illustration of these abstract concepts. Through this, readers gained insights into the constituent elements of the CAN stack and their interplay. Further enriching our discourse, we ventured into the nuanced features of both the PduR and COM modules. This entailed a deep dive into the communication mechanisms, highlighting how data is handled, routed, and transformed. Furthermore, the discussion on transfer property offered a granular view of how data transmission is modulated based on different conditions and requirements.

In summary, this chapter serves as a comprehensive guide for those looking to grasp the core tenets of AUTOSAR communication. The knowledge imparted herein is invaluable for readers as they embark on their AUTOSAR projects, offering both theoretical wisdom and practical insights to navigate the complexities of the framework effectively.

As we conclude this chapter, we will turn our attention to the next topic: the crypto stack in AUTOSAR. In the upcoming chapter, we will focus on exploring the crypto stack in detail and investigate how to secure communication with the communication stack through a use case.

Questions

1. What is the primary role of the COM module within the AUTOSAR framework?
2. How do signals fit into the larger framework of AUTOSAR communication?
3. What is the PduR module's function in the AUTOSAR communication paradigm?
4. In the context of the CAN stack example, can you highlight one key element and its significance?
5. What does the transfer property feature in the COM module refer to?
6. Why is the PduR module often referred to as a mediator in AUTOSAR communication?
7. How does the COM module manage the nuances of transmission and reception within AUTOSAR?
8. What are the different states for Nm?

Part 3: Beyond Fundamentals – Advanced AUTOSAR Concepts

This part addresses advanced topics within the AUTOSAR framework, such as cybersecurity and memory management. It explains the AUTOSAR Crypto Stack, **Secure Onboard Communication** (**SecOC**), and the **Non-Volatile Memory** (**NVM**) stack. The section concludes with a comprehensive use case, summarizing key concepts and encouraging further study and practical application. You will be prepared to tackle complex challenges in automotive software development and enhance your expertise in AUTOSAR.

This part has the following chapters:

- Chapter 8, *Securing the AUTOSAR System with Crypto and Security Stack*
- Chapter 9, *Dealing with Memory and Mode Management*
- Chapter 10, *Wrapping Up and Extending Knowledge with a Use Case*

8

Securing the AUTOSAR System with Crypto and Security Stack

In this chapter, we will embark on a journey through the aspects of automotive cybersecurity, beginning by providing an overview of the security landscape within the AUTOSAR framework.

Then we will explore the core components of the Crypto Stack Design, examining the principles and configurations for safeguarding sensitive data and ensuring secure communication between vehicle components. This introduction aims to get us familiarized with the concepts of security in automotive systems and demonstrate how AUTOSAR tackles these issues through its robust crypto stack. Let's embark on this journey together.

In this chapter, we will discuss the following:

- Introduction to automotive security
- Fundamentals of automotive security
- Security stack in AUTOSAR architecture
- Crypto Stack use cases and examples

Introduction to automotive security

Imagine driving a car that can be hacked remotely, jeopardizing not only your safety but also the security of your personal data. This nightmare scenario underscores the critical importance of automotive security in today's interconnected world.

The automotive industry has witnessed a remarkable transformation in recent years. Modern vehicles are equipped with advanced features, including infotainment systems, telematics, and autonomous driving capabilities. While these innovations have improved convenience and safety, they have also introduced new vulnerabilities. Just as we secure our homes and personal information, we must safeguard our vehicles against malicious actors seeking to exploit these vulnerabilities.

So, why is automotive security in the spotlight, especially in the context of AUTOSAR? Here are some points to consider:

- **Safety and lives at stake**: Beyond being a mode of transportation, vehicles are increasingly becoming computers on wheels. Any security breach can have life-threatening consequences. For example, imagine a scenario where a hacker gains control over a vehicle's braking system, causing a collision if the ECU is hijacked.
- **Data privacy**: Vehicles collect vast amounts of data, from driver preferences to geolocation data. Protecting this information is not just a matter of compliance with regulations such as GDPR; it's a matter of trust between consumers and automakers.
- **Financial implications**: Security breaches can result in significant financial losses, including recalls, legal liabilities, and damage to a company's reputation, as well as protection of the company IP.
- **Evolving threat landscape**: Hackers and cybercriminals continually develop new tactics and techniques to exploit vulnerabilities in automotive systems. As AUTOSAR evolves, so do the attack vectors. Staying ahead of these threats is crucial.
- **Regulatory compliance**: Governments worldwide are introducing stringent regulations and standards, mandating automotive cybersecurity. Non-compliance can result in severe penalties and market exclusion.

Now that you understand the importance of automotive security in today's interconnected world, let's explore the fundamental aspects of how this security is achieved within the AUTOSAR framework. In this section, we will explore the essential cryptographic mechanisms that safeguard data and communication in automotive systems.

Fundamentals of automotive security

In AUTOSAR, cryptographic mechanisms are indispensable for ensuring the security of data and communication within automotive systems. These mechanisms encompass both software and hardware cryptographic processing approaches, each offering distinct advantages and use cases. While software crypto relies on algorithms executed by the ECU's **central processing unit** (**CPU**) or microcontroller, hardware crypto leverages dedicated cryptographic modules or co-processors within the microcontroller or secure hardware components. These hardware modules, such as **hardware security modules** (**HSM**), are specifically designed to accelerate cryptographic operations, providing faster and more efficient processing. In AUTOSAR, cryptographic algorithms can be implemented using a combination of hardware and **software components** (**SWCs**), offering flexibility and tailored solutions to meet the diverse security requirements of automotive systems.

Figure 8.1 – Goals of security CIA

The primary objective of automotive security is to ensure the confidentiality, integrity, and availability of vehicle systems and data, which are discussed as follows:

- **Confidentiality** involves protecting sensitive information, such as user data and vehicle diagnostics, from unauthorized access and disclosure
- **Integrity** ensures that data and software remain accurate and trustworthy, preventing unauthorized modifications and tampering
- **Availability** guarantees that critical vehicle systems are operational and accessible when needed, defending against disruptions such as denial-of-service attacks

Additionally, robust authentication and authorization mechanisms are implemented to verify the identity of users and devices, granting appropriate access privileges based on their roles. These objectives collectively ensure that automotive systems are secure, reliable, and resilient against both accidental failures and malicious attacks.

Security in AUTOSAR

AUTOSAR recognizes the critical role of security in modern vehicles and has developed its set of security requirements and guidelines. These are designed to complement industry standards such as **ISO 21434** and **ISO 26262**, providing specific guidance for implementing security measures within the AUTOSAR architecture.

ISO 21434 is a comprehensive cybersecurity standard specifically designed for the automotive sector. It outlines the requirements for assessing and managing cybersecurity risks during the development and production of vehicles. ISO 21434 provides a structured framework for integrating security into the automotive development life cycle, ensuring that security is considered from the initial design phase through production and beyond.

ISO 26262 focuses on functional safety within the automotive industry. While it primarily addresses safety concerns, it also acknowledges the interplay between safety and security. A secure system can enhance safety by preventing malicious interference. ISO 26262 includes provisions for risk assessment related to cybersecurity, highlighting the importance of security considerations alongside safety.

Differentiating between safety and security

The two Ss, safety and security, are closely related terms in automotive software development. They often overlap, as a security breach or attack can have serious safety implications, such as compromising the functionality of critical systems such as brakes or steering.

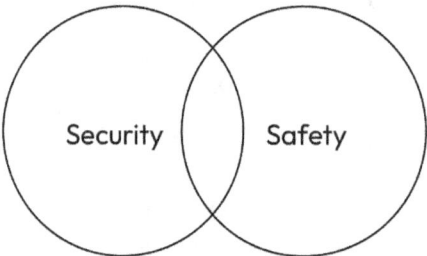

Figure 8.2 – Safety and security overlapping

Next, we will examine how AUTOSAR manages these intertwined aspects of safety and security.

Safety in AUTOSAR

Safety in AUTOSAR primarily concerns the prevention and mitigation of accidents and hazards. It encompasses mechanisms and processes that address physical failures and malfunctions in the vehicle's systems. Safety measures are designed to protect occupants, pedestrians, and the environment.

Security in AUTOSAR

In contrast, security in AUTOSAR is concerned with protecting the vehicle's systems from intentional attacks, unauthorized access, and data breaches. It encompasses measures to prevent cyber threats that can compromise the vehicle's functionality, privacy, and safety.

AUTOSAR security requirements cover a broad spectrum, including secure communication, cryptographic services, and access control. These requirements are designed to ensure that AUTOSAR-based systems can effectively counter various cyber threats. Rather than offering direct implementation guidelines, AUTOSAR provides building blocks to help secure ECU software. These building blocks support engineers in integrating robust security features, ensuring that automotive systems meet high-security standards and are resilient against potential cyber threats.

In the next section, we'll explore the BSW architecture of the crypto stack and its services within AUTOSAR, focusing on how it secures communication and data protection in automotive systems. We'll also examine the security stack's core components, highlighting its role in addressing the unique challenges of automotive security.

Security stack architecture in AUTOSAR

In this section, we will explore the security modules offered by AUTOSAR architecture, focusing on the modules that provide security functionality. AUTOSAR offers a toolbox of building blocks designed to help achieve security objectives in software. These tools enable engineers to integrate robust security features, ensuring that automotive systems meet high-security standards and are resilient against potential cyber threats.

The key security features in AUTOSAR are listed here:

- **Cryptographic services**: AUTOSAR provides a range of cryptographic services, including encryption, decryption, hashing, and digital signatures. These services are used for protecting data and ensuring its authenticity.

- **Access control**: AUTOSAR allows for fine-grained access control, specifying who can access specific functions, services, or data within the system. This enhances security by minimizing the attack surface.

- **Secure communication**: The architecture facilitates secure communication between ECUs, ensuring that data exchanged between components remains confidential and tamper-proof.

- **Secure boot**: AUTOSAR supports secure boot mechanisms, which help in verifying the integrity of SWCs during the boot-up process, safeguarding against unauthorized code execution.

> Note on AUTOSAR security features
>
> The security features provided by AUTOSAR are specified through detailed requirements, not directly as code implementations. These requirements outline the necessary security measures and functionalities that need to be integrated into automotive systems to ensure robust protection against cyber threats. For instance, AUTOSAR defines requirements for secure communication, cryptographic services, access control, and secure boot, among others. These requirements act as guidelines for engineers to design and implement security features in their ECU software, ensuring that the systems meet high-security standards and are resilient against potential cyber threats.

Cryptographic techniques

Cryptography forms the foundation of secure communication in AUTOSAR. The following cryptographic techniques are commonly used:

- **Encryption**: Data is encrypted before transmission, making it unreadable without the correct decryption key. This ensures data confidentiality, even if the data is intercepted by unauthorized entities.

- **Digital signatures**: Digital signatures are used to verify the authenticity and integrity of data. When data is signed, receivers can verify that it hasn't been altered during transmission and that it originated from a trusted source.

- **Hash functions**: Hash functions generate fixed-size values (**hashes**) based on input data. These hashes are used to verify data integrity. If the received hash matches the calculated hash of the received data, it's likely that the data has not been tampered with.

The following diagram, *Figure 8.3*, describes the BSW architecture the components of the crypto stack:

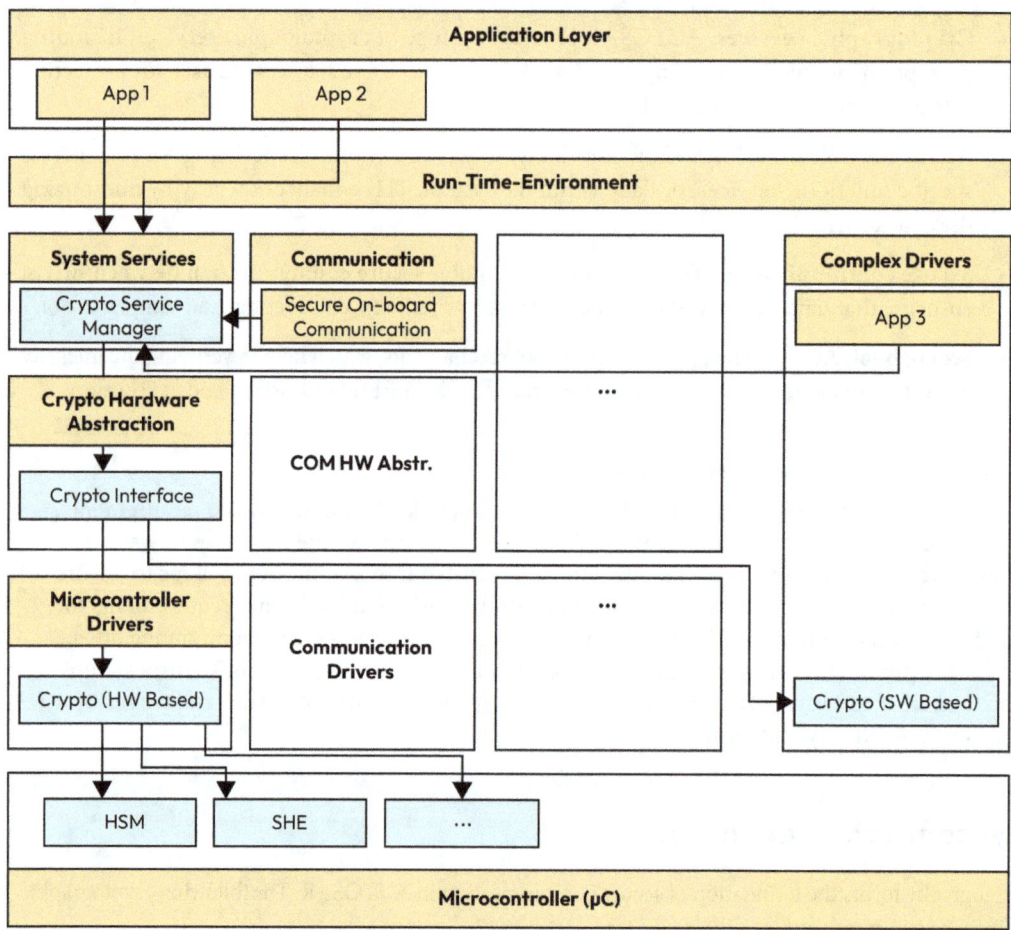

Figure 8.3 – Security Stack Architecture with in AUTOSAR

In AUTOSAR, the Crypto Stack is like a Swiss army knife that is used to ensure the security and integrity of data within automotive systems. It consists of several key modules, including the **Crypto Stack Manager (CSM)**, **Crypto Interface (CryptoIf)**, and the Crypto module. Each module serves a distinct purpose and contributes to the overall security architecture of AUTOSAR.

Crypto service manager module

The CSM is responsible for coordinating and managing cryptographic operations within the AUTOSAR system. Its primary functions include the following:

- **Request handling**: CSM receives cryptographic requests from various SWCs, services, or modules within the system. These requests can include operations such as encryption, decryption, hash generation, and digital signature creation or verification.
- **Resource management**: CSM manages cryptographic resources, such as cryptographic keys and algorithms. It ensures that keys are securely stored and that the appropriate cryptographic algorithms are available for use.
- **Key handling**: CSM oversees key management, including key generation, distribution, and revocation. It ensures that cryptographic keys are used securely and in compliance with security policies.
- **Error handling**: CSM handles cryptographic errors and exceptions. If a cryptographic operation fails or encounters an issue, CSM provides error codes and information for diagnostics and recovery.

Job concept in CSM

The **Crypto Service Manager (CSM)** in AUTOSAR acts as a service provider where users, such as SWCs and basic software modules such as **Secure Onboard Communication (SecOC)**, request cryptographic operations such as encryption and decryption. The Crypto Service Manager manages these requests by prioritizing them as *jobs* to ensure efficient processing.

A job refers to an instance of a static cryptographic operation that has been configured and could be submitted for execution. Each job involves specific parameters and operations related to cryptographic tasks such as encryption, decryption, **Message Authentication Code (MAC)** generation, and verification.

Here are some API examples:

- Csm_Encrypt:

  - **Description**: Encrypts data using a specified cryptographic algorithm
  - **Syntax**:

    ```
    Std_ReturnType Csm_Encrypt( uint32 jobId, Crypto_
    OperationModeType mode, const uint8* dataPtr, uint32 dataLength,
    uint8* macPtr, uint32* macLengthPtr );
    ```

- `Csm_MacGenerate`:
  - **Description**: This function is used to generate a MAC for the provided data
  - **Syntax**:
    ```
    Std_ReturnType Csm_MacGenerate ( uint32 jobId, Crypto_
    OperationModeType mode, const uint8* dataPtr, uint32 dataLength,
    uint8* macPtr, uint32* macLengthPtr );
    ```
- `Csm_Hash`:
  - **Description**: Computes the hash value of the given data
  - **Syntax**:
    ```
    Std_ReturnType Csm_Hash( uint32 jobId, Crypto_OperationModeType
    mode, const uint8* dataPtr, uint32 dataLength, uint8* resultPtr,
    uint32* resultLengthPtr);
    ```

> **Note**
>
> Most of the service APIs for encryption, decryption, hashing, and MAC calculation in the AUTOSAR CSM use similar parameters. These parameters, which may vary slightly depending on the specific service, typically include the following:
>
> `JobId`: Holds the identifier of the job using the Crypto Service Manager service
>
> - `Mode`: Indicates which operation mode(s) to perform
> - `DataPtr`: Pointer to the plaintext data to be encrypted
> - `data Length`: Contains the number of bytes to encrypt
> - `result LengthPtr`: Length of the result data pointer
> - `resultPtr`: Pointer to the buffer where the ciphertext will be stored
> - `KeyId`: Identifier of the key to be used for encryption
> - `Return Value`: Status of encryption (e.g., E_OK, E_NOT_OK, CRYPTO_E_BUSY, CRYPTO_E_KEY_NOT_VALID, etc.)

Crypto Service Manager process flow

The Cryptographic Service Manager in AUTOSAR manages cryptographic operations through a structured sequence of steps to ensure secure data handling. This process includes:

1. **Job submission**: The job is submitted to the Crypto Service Manager with all relevant parameters, such as SWC A, and would request `CSM_Encrypt`, referencing Job ID 1. This would include the key used for the encryption, as well as the cryptography primitive used.
2. **Validation**: The CSM validates the job parameters, including key validity and data length.

3. **Driver selection**: Based on the primitive type and key requirements, the CSM directs the job to the appropriate cryptographic driver.
4. **Execution**: The cryptographic driver processes the job according to the specified operation mode.
5. **Completion**: The result is returned, and the status of the job is updated (e.g., success or error).

The Crypto Service Manager in AUTOSAR acts as a service provider where users, such as SWCs and basic software modules such as SecOC, request cryptographic operations such as encryption and decryption. The CSM manages these requests by prioritizing them as *jobs* to ensure efficient processing.

The next layer of the crypto stack within the ECU abstraction layer would be the *CryptoIf*.

Crypto Interface module

The *CryptoIf* serves as an intermediary layer between the CSM and the actual crypto module responsible for performing cryptographic computations. Its key functions include the following:

- **Abstraction**: CryptoIf abstracts the underlying cryptographic functionality provided by the Crypto module. This abstraction simplifies the interaction between CSM and the crypto module, allowing for flexibility in choosing and changing cryptographic algorithms and providers.
- **Standardization**: CryptoIf standardizes the interface for cryptographic operations, making it easier to integrate different cryptographic libraries or HSMs into the AUTOSAR system.
- **Resource mapping**: CryptoIf maps cryptographic resources requested by CSM, such as keys and algorithms, to the corresponding resources provided by the crypto module.

Now it is time to move to the lowest layer in the stack. That's the Crypto Driver, which could be an SW or a hardware implementation as we are going to see later.

Crypto module

The crypto module is the core component responsible for performing cryptographic operations, such as encryption, decryption, hashing, digital signatures, and key management. Its importance lies in its ability to execute these operations securely and efficiently. Key functions of the Crypto module include the following:

- **Cryptographic computations**: The Crypto module implements cryptographic algorithms, ensuring that data is encrypted, decrypted, hashed, or signed as required.
- **Secure key storage**: The module securely stores cryptographic keys, protecting them from unauthorized access. It may interact with HSMs for enhanced key protection or store them in **Non-Volatile Memory (NVM)**.

The Crypto Stack, comprising the CSM, CryptoIf, and the crypto module, is needed for establishing and maintaining a secure communication and data protection environment within AUTOSAR systems. It allows for flexible integration of cryptographic functionality while ensuring that cryptographic operations are performed reliably and securely, safeguarding the confidentiality, integrity, and authenticity of data exchanged within the automotive ecosystem.

Since we are exploring the crypto driver, let's briefly discuss the different variations of the essence of the driver.

Hardware- and software-based implementations

Two primary approaches to cryptographic processing are software and hardware crypto, each with its distinct advantages and use cases.

Software crypto

Software cryptographic processing relies on algorithms executed by the vehicle's CPU or microcontroller. It offers flexibility and versatility, allowing cryptographic functions to be implemented in software libraries. Software crypto can adapt to different hardware platforms, making it a practical choice for systems with limited dedicated cryptographic hardware. However, it may consume more CPU resources and processing time compared to hardware-based solutions, potentially affecting overall system performance.

Hardware crypto

Hardware cryptographic processing, on the other hand, leverages dedicated cryptographic modules or co-processors within the microcontroller or secure hardware components. These hardware modules are specifically designed to accelerate cryptographic operations, providing faster and more efficient processing. Hardware crypto offers enhanced security by protecting cryptographic keys and operations from certain software-based attacks. It is particularly beneficial for performance-critical applications and systems with stringent security requirements.

One notable feature within AUTOSAR is the ability to use multiple cryptographic modules, both software and hardware-based, and configure them as needed. Through configuration, users can determine which cryptographic module manages specific cryptographic primitives or operations. This flexibility allows AUTOSAR developers to tailor the security architecture to their system's unique requirements, optimizing the balance between performance, security, and resource utilization. This capability empowers users to make informed decisions regarding the allocation of cryptographic tasks, ensuring that cryptographic operations are performed efficiently and securely within the AUTOSAR framework.

Figure 8.4 shows a block diagram of a secure system architecture. The **Secure Space** encloses critical security components, ensuring that sensitive operations and data are protected. The other components are part of the general system architecture, interfacing with the secure space and handling standard operations.

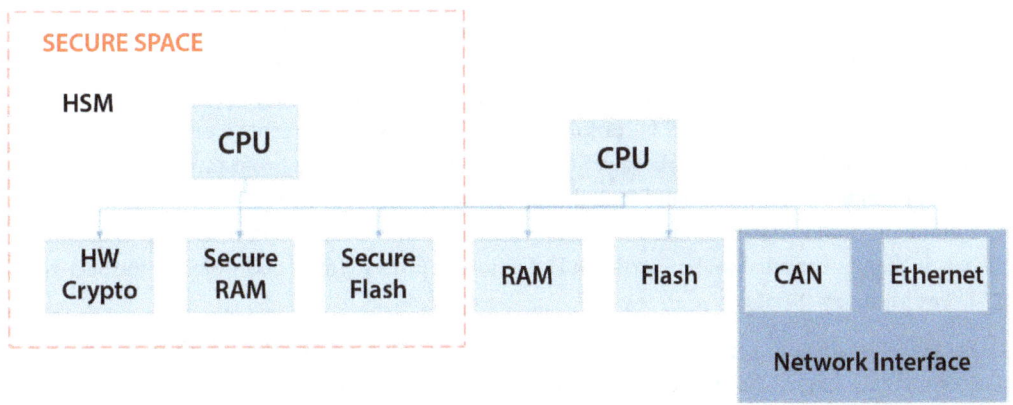

Figure 8.4 – Secure system architecture block diagram

Utilizing a hardware component such as an HSM offers a significant edge over a solely software-based approach because it can establish a dedicated, highly secure environment encompassing essential elements such as the core, secure memory, and hardware accelerators. Within this secure environment, several key functions are facilitated:

- Provision of security services to applications
- Safe storage of keys and associated security assets.
- Efficient management of keys and certificates
- Execution of cryptographic algorithms, employing both hardware and software implementations to enhance performance and security

Now that we've unraveled the core crypto stack modules, perhaps it's time we scratch the surface of some of the other helper modules, as well as their description that are used with the crypto modules to enhance the overall ECU security.

Other helper modules to achieve security

Some of the helper modules we will discuss are located in other stacks (for example, IdsM and KeyM) are part of the Crypto stack. However, SecOC is part of the COM stack as it handles communication, yet it works closely with the Crypto Stack to achieve secure PDUs communication:

- **Key Management (KeyM)**: This is responsible for handling cryptographic keys. It ensures the secure generation, storage, distribution, and life cycle management of keys used for various cryptographic operations.

- **Intrusion Detection System Manager (IdsM)**: The **Intrusion Detection System (IDS)** functions as a guard in a vehicle, collecting data from security sensors distributed throughout the vehicle and sending it to a central system for detailed analysis. This analysis helps in formulating robust future security measures. IDS is composed of several key components working together to detect and respond to cyber threats. Software sensors, which are programmed algorithms, detect potential cyber threats and send this information to the IdsM. The IdsM acts as a filtering system, processing the incoming data to identify and isolate relevant information. This qualified data is then transmitted to the **Intrusion Detection Reporter (IdsR)** and stored in the **Security Event Memory (SEM)**.

- **SecOC**: This ensures secure data transmission within vehicle networks. It integrates with the Crypto Stack to provide cryptographic protection, safeguarding data integrity and authenticity as part of the communication stack.

- **Diagnostic Communication Manager (DCM)**: This uses the Crypto Stack to manage security access levels by performing cryptographic operations, such as authentication and key verification. When a diagnostic session requires access to restricted functions, the DCM relies on the Crypto Stack to securely handle authentication challenges and responses, ensuring only authorized tools or users can access sensitive ECU operations.

Now that we have highlighted the importance of security for automotive ECUs, let's explore some common security use cases and examine how the crypto stack is used to achieve these use cases. This exploration will help us understand the practical applications and integration of cryptographic services in enhancing automotive security.

Crypto Stack use cases and examples

The Crypto Stack in AUTOSAR is a critical component that serves various use cases for securing data and communication within automotive systems. Some of the key use cases for the Crypto Stack in AUTOSAR include secure communication, data encryption, and digital signatures.

Data encryption

Let's consider a scenario. In an AUTOSAR-based embedded application, **SWC A** needs to securely transmit sensitive data to **SWC B** over a **Controller Area Network (CAN)** bus. To ensure data confidentiality, SWC-A employs data encryption using the AUTOSAR Crypto Stack.

Here's a list of the actors:

- **SWC-A**: The sender of sensitive data
- **SWC-B**: The recipient of the encrypted data

- **AUTOSAR Crypto Stack**: The cryptographic module responsible for data encryption and decryption
- **Security Key Management**: The component responsible for managing cryptographic keys

Here are the preconditions:

- SWC-A and SWC-B are configured and operational within the AUTOSAR-based system
- The AUTOSAR Crypto Stack is properly configured and integrated into the system
- Encryption keys, including a shared secret key for SWC-A and SWC-B, have been securely established and stored by the Security Key Management component

Figure 8.5 shows the data exchange process between SWC A and SWC B, located on two different ECUs.

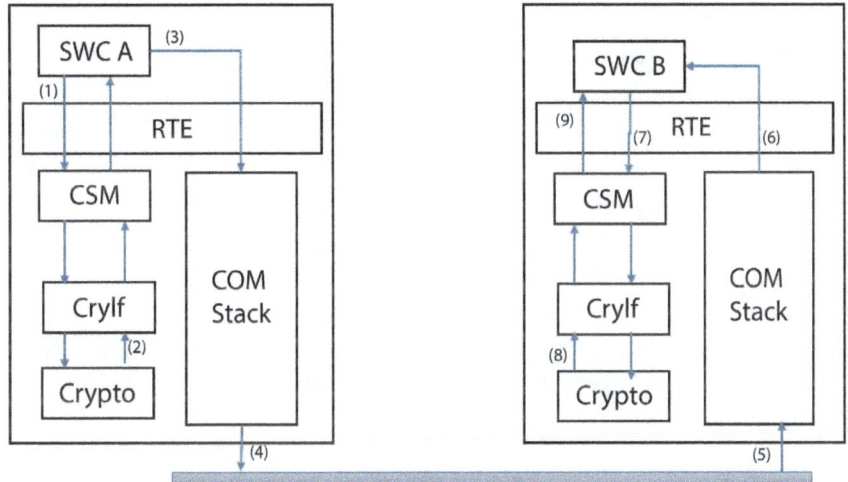

Figure 8.5 – Encrypted data shared between two ECUs

Here's the sequence for the main flow:

1. **SWC-A data preparation**: SWC-A generates sensitive data that needs to be transmitted to SWC-B. Before sending the data, SWC-A initiates the data encryption process using the AUTOSAR Crypto Stack.
2. **Request for encryption**: SWC-A sends a request to the AUTOSAR crypto service manager to encrypt the data for secure transmission. The request specifies the data to be encrypted and indicates that a specific key should be used for encryption, as shown in **step 1** within *Figure 8.5*.
3. **Crypto Stack encryption**: The AUTOSAR Crypto Stack receives the encryption request and uses the shared secret key to perform the encryption process.

4. **Encrypted data transmission**: The Crypto Stack returns the encrypted data to SWC-A, as shown in step 2 within *Figure 8.5*. Only then can SWC-A securely transmit the encrypted data over the CAN bus to SWC-B using the communication stack shown in **step 3**.

5. **Data reception by SWC-B**: SWC-B receives the encrypted data from SWC-A over the CAN bus. Since SWC-B is aware of the shared secret key, it is prepared to decrypt the incoming data. This is shown in **step 5** and **step 6** in *Figure 8.5*.

6. **Data decryption request**: SWC-B sends a request to the AUTOSAR Crypto Stack, indicating that it needs to decrypt the received data using the shared secret key, shown in **step 7**.

7. **Crypto Stack decryption**: The AUTOSAR Crypto Stack processes SWC-B's decryption request, utilizing the shared secret key to perform the decryption operation. It reverses the encryption process, restoring the original sensitive data.

8. **Decrypted data usage**: SWC-B now has access to the decrypted sensitive data, which can be processed and utilized as required for its intended purpose within the embedded system. This is shown in **step 8** and **step 9**.

Here are the post-conditions:

- SWC-A successfully transmitted sensitive data to SWC-B in an encrypted form, ensuring data confidentiality during transmission
- SWC-B received and decrypted the data, allowing it to access and utilize the original sensitive information

There's one exception. If the encryption or decryption process fails due to key-related issues or cryptographic errors, appropriate error-handling mechanisms within the AUTOSAR Crypto Stack are triggered to ensure the security and integrity of the data.

Here are some important notes:

- The use of a shared secret key for encryption and decryption between SWC-A and SWC-B highlights the importance of secure key management within the embedded AUTOSAR system
- The AUTOSAR Crypto Stack abstracts the low-level cryptographic operations, simplifying the implementation of data encryption and decryption for SWCs while maintaining a high level of security

Expanding on the previous example, we could then explore secure communication. This refers to the mechanisms and protocols used to ensure the integrity, authenticity, and confidentiality of data exchanged between various **electronic control units** (**ECUs**) within a vehicle's network. The primary goals are to prevent unauthorized access, data manipulation, and eavesdropping.

Secure communication

Secure On-Board Communication (SecOC) in AUTOSAR is a critical component that ensures the confidentiality and integrity of data exchanged between ECUs within a vehicle's network. It adds security features such as MACs and freshness values to **Protocol Data Units (PDUs)** before transmission. Let's explore how SecOC works in the context of a CAN network between two ECUs, ECU 1 and ECU 2, with an example.

Figure 8.6 illustrates how secure communication is managed between two ECUs using MAC. It will depict the interaction between the communication stack, the crypto stack, and the SecOc module to achieve this process.

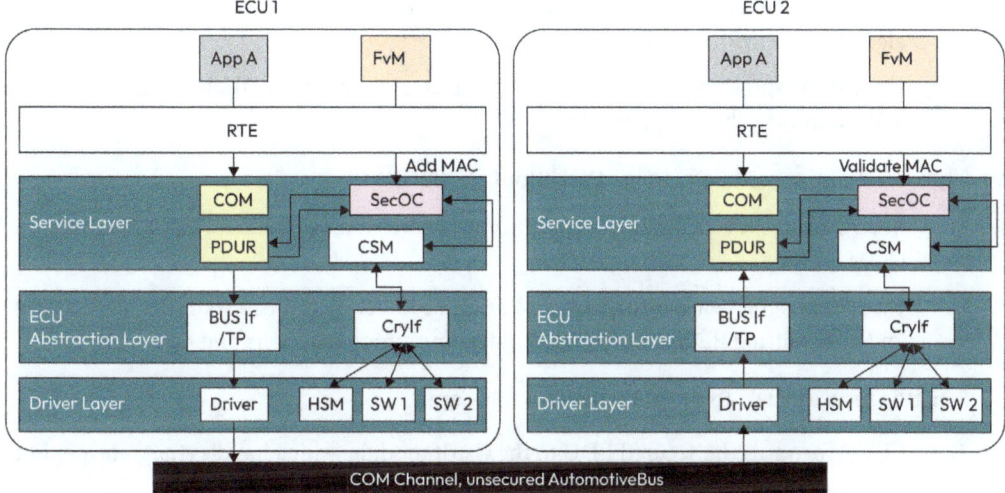

Figure 8.6 – SecOc

The following steps show the flow from ECU 1, where data is being secured with MAC, and how the data is verified in ECU 2:

1. Start by sending a PDU from SWC 1 in ECU 1:

    I. SWC 1 in ECU 1 generates a PDU that needs to be sent to ECU 2 over the CAN network. This PDU contains critical information, such as sensor data, that must be kept confidential and protected from tampering.

    II. The PDU is handed over to the **Communication (COM)** module within ECU 1. COM manages the transmission of PDUs between different ECUs.

2. Next, secure the PDU with SecOC:
    I. The PDU is then passed from COM to the SecOC module within ECU 1. SecOC is responsible for adding security features to the PDU.
    II. SecOC interacts with the CSM to initiate the security process. CSM manages cryptographic operations in AUTOSAR.
    III. CSM, in turn, communicates with the CryptoIf module to request cryptographic services.
    IV. CryptoIf acts as an intermediary between CSM and the crypto module, abstracting the cryptographic functionality.
    V. The crypto module is responsible for performing the actual cryptographic operations, including MAC calculation and freshness value generation, using the cryptographic algorithms provided by the crypto stack.
    VI. Once the cryptographic operations are complete, the secured PDU, including the MAC and freshness value, is returned to SecOC.
3. Then, send the secured PDU to the **PDU Router (PDUR)**:
    I. SecOC passes the secured PDU, including the MAC and freshness value, back to the COM module.
    II. COM sends the secured PDU to the **PDUR** module. PDUR is responsible for routing and transmitting PDUs between different communication layers and buses.
4. Send the secured PDU to the CAN network via CanIf:
    I. PDUR forwards the secured PDU to the CanIf module, which handles the transmission of data over the CAN network.
5. Now, it's time for receiving and verifying the secured PDU in ECU 2:
    I. ECU 2 receives the secured PDU via the CAN network through its CanIf module.
    II. The secured PDU is then handed over to ECU 2's SecOC module.
6. Verify and decode the secured PDU with SecOC:
    I. ECU 2's SecOC module interacts with the CSM, CryptoIf, and crypto modules, similar to ECU 1, to verify the MAC and freshness value and ensure data integrity and authenticity.
    II. If the MAC and freshness value checks pass, the secured PDU is considered authentic and trustworthy.
7. Pass the decoded PDU to SWC 2 in ECU 2:
    I. The decoded and verified PDU is then passed to ECU 2's COM module.
    II. COM forwards the PDU to the intended SWC 2 in ECU 2.

In this detailed process, we could examine how the CSM, CryptoIf, and the Crypto module secure and verify PDUs. Additionally, the PDUR is involved in routing the secured PDU to the CAN network.

Secure diagnostics

Secure diagnostics has become a key process in the maintenance and operation of modern vehicles. As cars evolve into highly connected and software-driven systems, the need to protect diagnostic processes from unauthorized access and cyber threats is more important than ever. It ensures that only authorized personnel can access and modify critical vehicle data.

The process begins with the technician logging into a diagnostic tool using secure credentials, which are then verified by the ECU. Once authenticated, the technician securely communicates with the ECU to retrieve diagnostic data or perform necessary actions, such as software updates. Throughout the process, all data transmission is encrypted to prevent tampering or interception. The validation and verification of the keys or access control is handled through DCM module and the crypto stack services. *Figure 8.7* shows a very brief sequence diagram on what happens in this process.

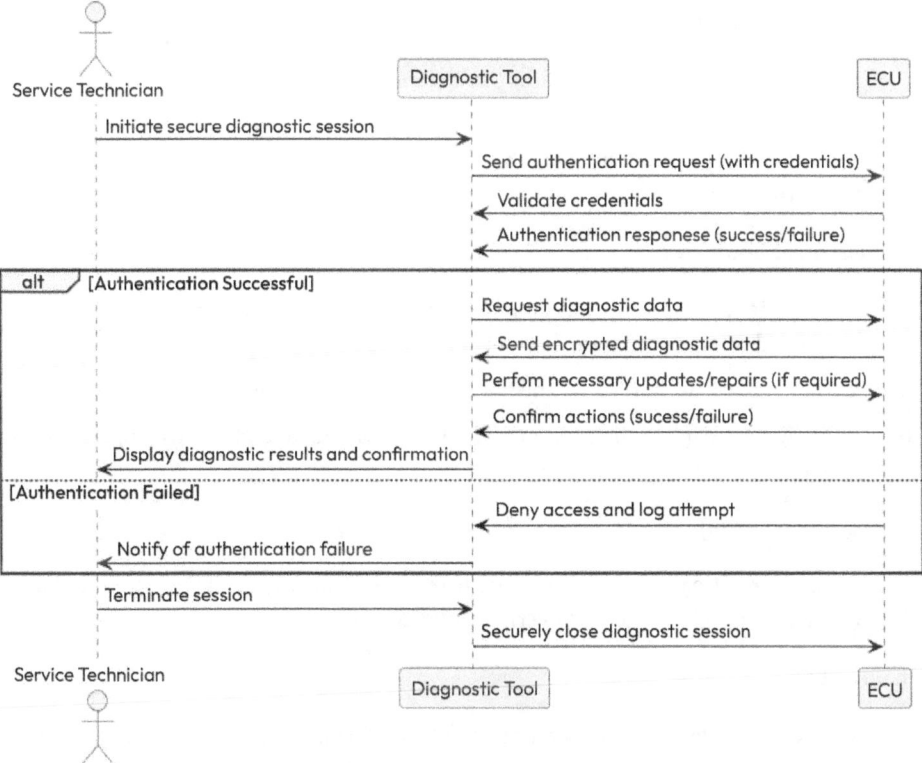

Figure 8.7 – Secure diagnostics process

Next, let's explore the concept of secure boot.

Secure boot in AUTOSAR

We'll start with an overview. **Secure boot** is a critical security mechanism employed in AUTOSAR-based automotive systems to ensure the integrity and authenticity of SWCs during the boot-up process. It verifies that only trusted and unaltered software is loaded and executed on ECUs within the vehicle. Secure Boot relies on cryptographic techniques, and the AUTOSAR Crypto Stack plays a pivotal role in enabling and enhancing the security of this process.

We'll consider a scenario. Let's explore a use case where Secure Boot is applied to ensure the integrity of the ECU software in an AUTOSAR-based vehicle.

Here are the steps to make that happen:

1. **Initialization**:
    I. The vehicle is powered on, and the ECU initiates the boot-up process.
    II. At this stage, the ECU's hardware performs a minimal self-test to ensure its functionality.

2. **Secure boot sequence**:
    I. Secure boot begins with the loading of the initial bootloader, which is stored in a **read-only memory (ROM)** of the ECU.
    II. A piece of code either within HSM or in a secure area called Root of Trust would try and verify the bootloader.

3. **Verification of bootloader**:
    I. The bootloader contains a digital signature, generated during its development and embedded when flashing the bootloader into an ECU.
    II. Crypto Stack is used to verify the bootloader's digital signature.
    III. If the signature verification fails, the boot process is halted, preventing the execution of unauthorized or tampered code.

4. **Loading the main application**:
    I. Once the bootloader's integrity is confirmed, it loads the AUTOSAR-compliant SW, which may also be signed and verified similarly as before.

5. **Chain of trust**:
    I. Secure boot establishes a chain of trust by ensuring that each loaded component is trusted and unaltered.
    II. If any signature verification fails, the loading process is halted, and the compromised component is not executed.

6. **Secure execution**:

   I. Once all SWCs have been verified and loaded, the ECU proceeds with the secure execution of the automotive applications.

Figure 8.8 illustrates an example of how a secure boot process can be achieved using cryptographic functions within the AUTOSAR framework.

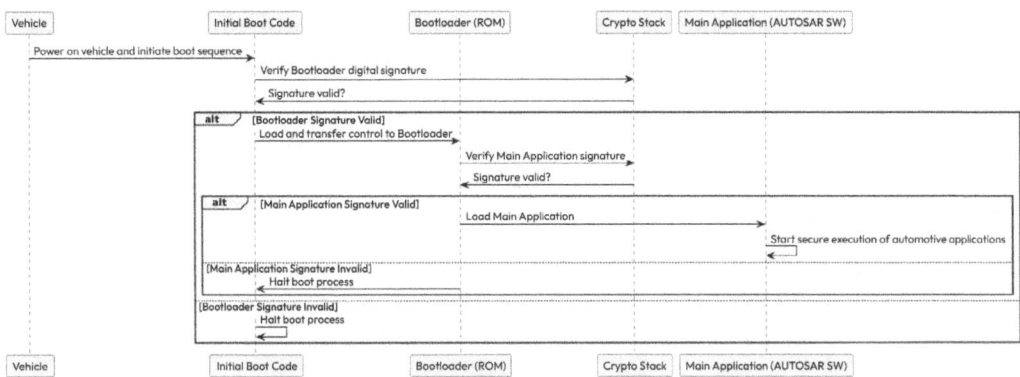

Figure 8.8 – Secure Boot process

Next, let's look at the role of Crypto Stack. The AUTOSAR Crypto Stack is instrumental in enhancing the security of the Secure Boot process:

- **Key management**: The Crypto Stack manages cryptographic keys used for verifying digital signatures, ensuring their confidentiality and integrity

- **Signature verification**: It performs signature verification of bootloaders, operating systems, and SWCs using the stored public keys

- **Tamper detection**: The Crypto Stack can also assist in detecting tampering attempts by monitoring the integrity of the SWCs and their signatures

- **Secure storage**: Cryptographic assets, including keys and certificates, are securely stored and accessed by the Crypto Stack to prevent unauthorized manipulation

In this use case, the combination of Secure Boot and the AUTOSAR Crypto Stack ensures that only trusted and unaltered software is executed, safeguarding the ECU's functionality, reliability, and overall vehicle security. It provides a robust defense against unauthorized code execution and potential security breaches.

Summary

In this chapter, the focus was on automotive cybersecurity within the AUTOSAR framework. It began by highlighting the significant importance of automotive security in our interconnected world, emphasizing the potential risks associated with remote hacking and data breaches that can impact safety, data privacy, financial stability, and regulatory compliance. The chapter delved into the fundamental principles of automotive security, defining its objectives which include confidentiality, integrity, availability, and authentication. It also made a clear distinction between safety and security, underlining the necessity of addressing both aspects within the context of AUTOSAR.

Furthermore, the chapter underscored the critical role played by security standards in the automotive industry, outlining key standards such as ISO 21434, ISO 26262, and AUTOSAR's own security requirements and guidelines. It emphasizes the need for adherence to these standards to ensure uniformity and regulatory compliance across the industry.

The chapter proceeded to explore the security features within the AUTOSAR architecture, with a particular focus on the Crypto Stack. It elaborated on the core components of the Crypto Stack, including the CSM, CryptoIf, and the crypto module, elucidating their roles in managing cryptographic operations securely.

SecOC was highlighted as a critical component ensuring data confidentiality and integrity between ECUs within a vehicle's network. The chapter offered a detailed explanation of the steps involved in securing and verifying data transmission.

In conclusion, the chapter emphasized the importance of the information presented for developers and newcomers to AUTOSAR. It stressed that security is not only about safeguarding data and systems but also about protecting lives, financial stability, and complying with regulatory requirements. Developers were encouraged to adopt secure coding practices, recognize the existence of legacy systems, and consider supply chain security as an integral component of an effective automotive security strategy.

In the next chapter, we shall explore the Memory and Mode management modules

Questions

1. Why is automotive security crucial within the AUTOSAR framework?
2. What are the fundamental objectives of automotive security?
3. How does safety differ from security in the context of AUTOSAR?
4. What is the role of the Crypto Stack within AUTOSAR, and what are its core components?
5. How does SecOC contribute to automotive security within AUTOSAR?

9

Dealing with Memory and Mode Management

In the ecosystem of AUTOSAR, where reliability, performance, and safety are paramount, the management of **non-volatile memory** (**NvM**) and the **Basic Software Mode Manager** (**BSWM**) emerge as a critical aspect of software development. This chapter dives into the mechanisms by which **software components** (**SWCs**) leverage NVM to ensure persistent storage of crucial data, configurations, and diagnostic information, while also exploring the role of the BSWM in managing the operational modes of basic software (**BSW**) modules.

The discussion begins by exploring the fundamental concepts of NVM and its significance within the AUTOSAR framework. We highlight the architecture and functionalities of the memory stack, emphasizing its role in managing data persistence and integrity. Our exploration extends to understanding the design and implementation of NVM operations within SWCs, detailing the memory stack modules.

Simultaneously, we introduce the concept of the BSWM, explaining its importance in coordinating the states and modes of various BSW modules. We delve into the mechanisms of the BSWM, exploring how it manages mode transitions, coordinates with other modules, and ensures a coherent operational state across the system.

In this chapter, we will look at the following topics:

- Memory stack in AUTOSAR
- Mode management

Memory stack in AUTOSAR

Before delving into NVM, it's important to understand the concept of a memory stack and how NVM fits into this context within the AUTOSAR framework. Imagine the memory stack as a library's storage system, where each book represents a piece of data that needs to be stored, cataloged, and retrieved efficiently. In this analogy, the memory stack in AUTOSAR is the librarian that manages this system, ensuring that essential information is preserved across power cycles and system resets.

Just as a librarian ensures that important books are safely stored and readily accessible, the memory stack in AUTOSAR manages data storage and retrieval. It ensures that critical information remains intact even when the system is turned off or restarted. This section will provide an overview of the memory stack, introduce the significance of NVM, and explain its integration within the memory stack to enhance system reliability and data integrity. We will explore the structure and functionality of the memory stack, detailing its layers and modules, and then discuss the different types of NVM blocks and their specific use cases. By the end of this section, you will have a comprehensive understanding of the memory stack's architecture, its components, and how it ensures data availability and integrity in automotive systems.

Significance of non-volatile memory in AUTOSAR

The memory stack safeguards vital data and configurations throughout power cycles and system resets. In an **electronic control unit** (ECU), it is usually needed to preserve critical information, such as vehicle settings, calibration parameters, **diagnostic trouble codes** (DTCs), and other indispensable data necessary for optimal operation. Similarly, the need to retrieve stored information persists after power cycles. To ensure seamless storage and maintenance of such data, the BSW incorporates an NVM stack, which can utilize **electrically erasable programmable read-only memory** (EEPROM) or flash memory for storage. Certain configurable settings control how and when data is written to and read from memory. These settings determine whether data is written immediately after it changes or deferred until the system shuts down and whether data is automatically retrieved at startup or only when specifically requested. The NVM stack manages these configurations to ensure data integrity and availability, making sure that critical information is securely stored and accessible when needed.

Understanding the memory stack

The memory stack in AUTOSAR consists of several layers and modules, each serving a specific purpose in managing non-volatile data storage and retrieval. *Figure 9.1* illustrates the architecture of the memory stack in an AUTOSAR ECU, showing the interaction between different layers and components involved in managing NVM:

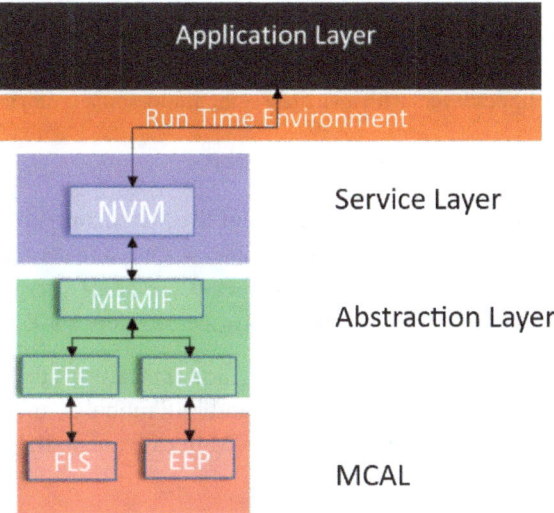

Figure 9.1 – Memory stack architecture

In the following subsections, we will discuss the functionality of each module within a layer in brief.

Service layer

Within the service layer, the **non-volatile random access memory (NVRAM) Manager** module serves as the core component of the memory stack in AUTOSAR. It provides standardized interfaces and services for managing NVM operations within the AUTOSAR framework.

Key functionalities of this layer include data block management, read and write operations, and error handling.

The NVM module acts as an interface between upper-layer SWCs and the underlying hardware, abstracting the complexities of NVM management.

Abstraction layer

In embedded devices, data storage can occur in emulated flash or an external EEPROM. The **Flash EEPROM Emulation** (FEE) and **EEPROM Abstraction** (EA) modules in AUTOSAR serve to abstract the lower storage devices, offering a standardized interface for managing data storage in embedded devices. These modules not only manage read/write/erase operations but also ensure reliability and efficient utilization of storage resources. Both provide interfaces for read, write, erase, and block management tasks and wear leveling.

The **Memory Abstraction Interface (MemIf)** abstracts the underlying FEE and EA modules, acting as a routing layer to their functions as it shall provide an abstraction from the number of underlying FEE or EA modules and provide upper layers with a virtual segmentation on a uniform linear address space.

MCAL

EEPROM and flash memory are integral components of the **Microcontroller Abstraction Layer (MCAL)**. The EEPROM driver provides services for reading, writing, and erasing data blocks. Internal drivers directly access microcontroller hardware within the MCAL, while external drivers use handlers (typically **serial peripheral interface (SPI)**). Both types have identical functional requirements and APIs.

Figure 9.2 illustrates the interaction between the memory stack modules on the flow of write operations from the NVM module through the NVM and FEE modules, and ultimately to the FLS module, which handles the actual writing process to NVM in case the underlying physical memory is emulated flash memory.

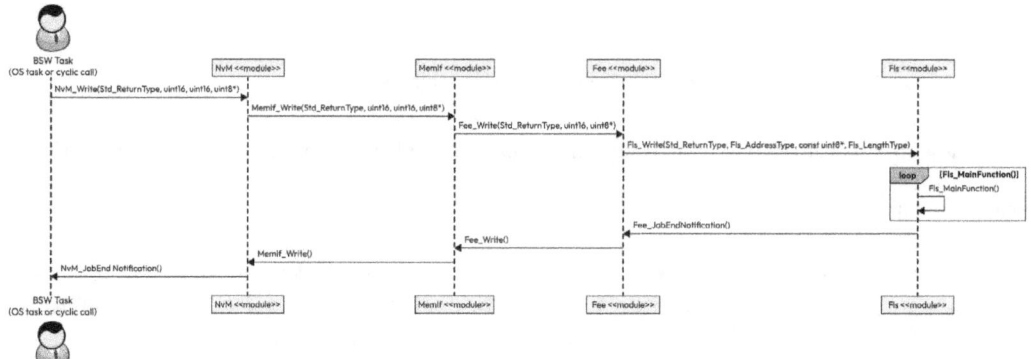

Figure 9.2 – Memory stack architecture

Next, we will explore the storage object types of NVM, examining their roles and how they contribute to the realization of the functionality of the NVM operations.

Storage objects

The NvmM module defines fundamental storage units known as **blocks** or **NVRAM blocks**. These blocks act like individual compartments in a storage facility where your NVM data is securely stored and organized. Just as each compartment in a storage facility keeps your items safe and easily retrievable, these blocks ensure that your data remains intact and accessible even when the power is turned off. *Figure 9.3* defines the basic storage objects.

Memory stack in AUTOSAR 187

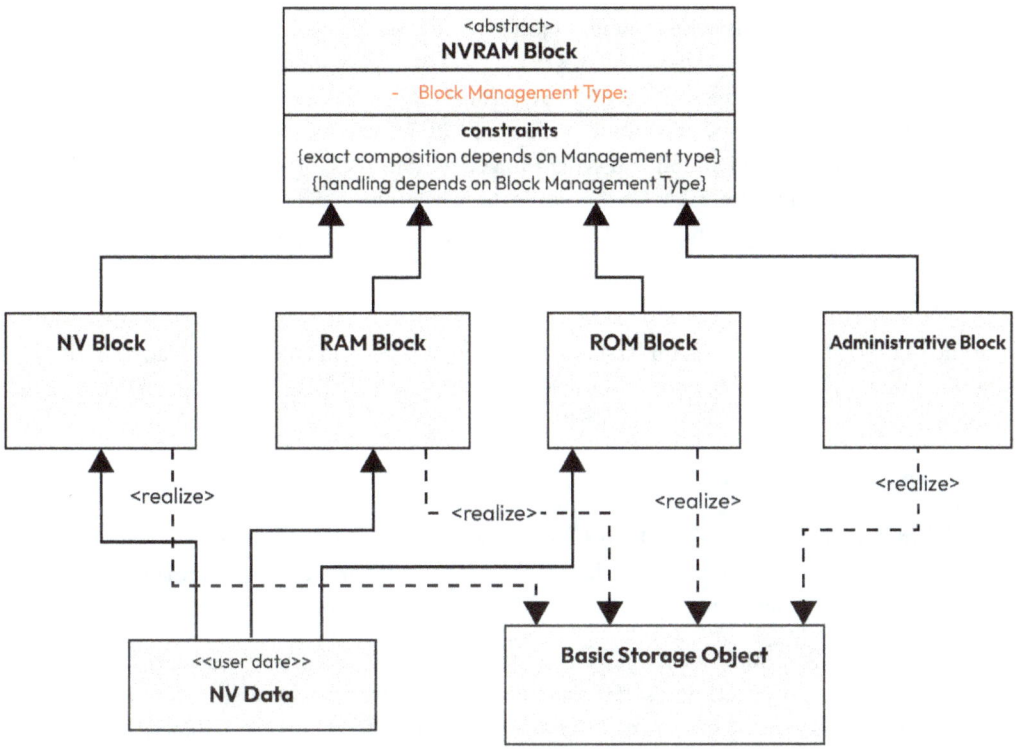

Figure 9.3 – Memory stack architecture

The NvmM module's component model defines four types of basic storage objects, which constitute the NVRAM block management types: the NV block, RAM block, ROM block, and Administrative block. Each of these blocks serves a specific function and collectively realizes the basic storage objects. We will discuss these in brief:

- **RAM block**: This provides temporary storage for active data, to be accessed/used directly by a user example Application.
- **NV block**: This handles non-volatile data storage, in the physical device flash or EEPROM.
- **ROM block**: This provides default data for the NV block as provided by the application, in case an error occurs reading a block.
- **Administrative block**: This manages metadata and control information, such as block ID, block size, and so on. It is not accessible to any other module outside the NvmM.

In the preceding section, we discussed storage objects. These block types can be combined in different configurations to create various types of NVRAM blocks. In the following section, we will cover the various block management types in NVRAM. A typical NVRAM block comprises one or more NV blocks that are allocated in physical storage, such as EEPROM or flash memory, an optional RAM block associated with the application and synchronized with the NV block, and an optional ROM block to store default data. Understanding these blocks is helpful for efficient memory management and ensuring the integrity and longevity of stored data in automotive applications.

Block types

By the end of this section, you will understand the characteristics and use cases of Native NVRAM blocks, Redundant NVRAM blocks, and Dataset NVRAM blocks, along with their importance in maintaining data integrity and system reliability.

Native NVRAM block type

This is the simplest form of an NVRAM block, consisting of a single instance of an NV block, a RAM block, a ROM block, and an Administrative block. This block stores non-critical information that does not require redundancy, data integrity is not as critical, and data can be easily reconstructed if lost.

Use case: Consider an automotive ECU for a vehicle infotainment system storing configuration parameters that control user settings, such as audio volume, display brightness, and preferred radio stations. These settings are easy to recreate or reset to default if lost, making native blocks suitable for this purpose due to their simplicity and minimal overhead.

Redundant NVRAM block type

Redundant blocks provide an additional layer of data protection by maintaining multiple copies of the same data. Thus, two NV blocks are used to realize this type. This redundancy ensures that even if one copy becomes corrupted or lost, the system can still retrieve the data from another copy.

Use case: In safety-critical systems, such as an **electronic brake system** (**EBS**), where data integrity is paramount, redundant block management is required. Two copies of the NV block are maintained to ensure redundancy. For instance, if one copy of the brake calibration data becomes corrupted due to a fault, the EBS ECU can switch to the redundant copy, thereby maintaining continuous operation without compromising safety.

Dataset NVRAM block type

Dataset blocks are more complex and involve managing multiple versions of the data. These blocks allow the storage of different datasets within the same NVRAM block, each representing a different version or state of the data. This is particularly useful for applications that need to keep track of historical data or configurations, providing the ability to revert to previous states or compare different datasets.

Use case: In an automotive climate control system, a dataset block management approach can be employed to store personalized air conditioning temperature settings for different users of the car. Each user can have their preferred temperature settings saved as a separate dataset. When a user enters the car and selects their profile, the climate control system retrieves and applies their preset temperature from the corresponding dataset block.

> **Managing block types**
>
> Managing block types effectively is crucial for extending the EEPROM lifespan and ensuring data safety. Since each memory location has physically limited write cycles, it's important to choose block types wisely based on write frequency and data importance. Strategic selection of block types helps maintain data integrity and maximizes memory longevity.

After discussing the various types of NVRAM blocks, it is important to understand how these blocks are accessed.

Block properties

We will define how these blocks can be read and written, focusing on two of the most common parameters that can be configured: `NvMSelectBlockForReadAll` and `NvMSelectBlockForWriteAll`:

- `NvMSelectBlockForReadAll`: The `read_all` flag, when enabled, indicates whether the block can be included in a `read all` operation, where all NVM blocks are read sequentially. This is useful for initializing the system state by reading all necessary configuration data at once.

  Behavior: If the `read_all` flag is enabled for a block, the NVM module will read all data stored within that block:

- `NvMSelectBlockForWriteAll`: The `write_all` flag, when enabled, indicates whether the block can be included in a `write all` operation, where all NVM blocks are written sequentially. This is beneficial for saving the current state of the system by writing all relevant data to NVM at once usually during shutdown.

- `NvMNvBlockLength`: This is a configuration parameter that defines the size of an NVM block in bytes. This parameter specifies how much data the block can store, determining the memory allocation for that block within the NVM. The size of the block is crucial for ensuring that it can hold all necessary information and align with the memory layout and management strategies of the system.

> **NvM Block Descriptor configuration parameters**
>
> For more detailed information about the configuration parameters, you can refer to the NvM AUTOSAR specifications, such as *release R20-11*, specifically *section 10.2.3 NvMBlockDescriptor*. This section provides comprehensive insights into the various parameters and their configurations.

In the following table, we will discuss the API provided by NvmM for accessing an NV Block:

API	Description
`NvM_ReadBlock`	Reads data from a specified data block in NVM.
`NvM_WriteBlock`	Writes data to a specified data block in NVM.
`NvM_SetRamBlockStatus`	Sets the status of a RAM block, indicating whether it is valid.
`NvM_GetErrorStatus`	Retrieves the error status of a specified block.
`NvM_ReadAll`	Request for reading all for all blocks marked for "read all".
`NvM_WriteAll`	Request for writing all data marked for "write all".

Table 9.1 – NvmM API functions and descriptions

In the next section, let's explore an example of how the memory stack is utilized in an ECU.

Use case

Scenario: In an automotive ECU responsible for engine control, there's a crucial need to store calibration parameters persistently across power cycles. These parameters include fuel injection timing, ignition timing, and air-fuel ratio settings, which directly impact engine performance and emissions control. The ECU interacts with a SWC, let's call it `EngineControlSWC`, to manage these calibration parameters. The SWC utilizes functions provided by the NVM module to read and write calibration data.

NVM configuration: The NVM module is configured with multiple data blocks, each serving a specific purpose. For this use case, let's consider a native block management approach. There's one NVM block dedicated to storing all calibration parameters. Additionally, the `read_all` flag is selected for this block, ensuring that all calibration parameters are read during startup. Similarly, the `write_all` flag is chosen, indicating that all blocks marked for `write_all` will be written during shutdown.

Startup sequence: Upon system startup, the following sequence of events takes place:

1. `EngineControlSWC` initializes its data structures and registers with the NVM module.
2. `EngineControlSWC` requests the NVM module to read all calibration parameters stored in the designated data block.
3. Since the `read_all` flag is selected, the NVM module reads all calibration parameters from the block and provides them to `EngineControlSWC`.
4. `EngineControlSWC` utilizes the retrieved calibration parameters for engine control algorithms and initialization routines.

Shutdown sequence: During system shutdown, the following sequence of events takes place:

1. EngineControlSWC finalizes its operations and prepares for shutdown.
2. EngineControlSWC requests the NVM module to write all calibration parameters back to the designated data block.
3. As the write_all flag is chosen, the NVM module writes all calibration parameters to the block, ensuring that all changes made during runtime are persistently stored.
4. The NVM module confirms the successful completion of the write operation, allowing EngineControlSWC to proceed with the shutdown process.

Figure 9.4 shall show how this process or flow shall go.

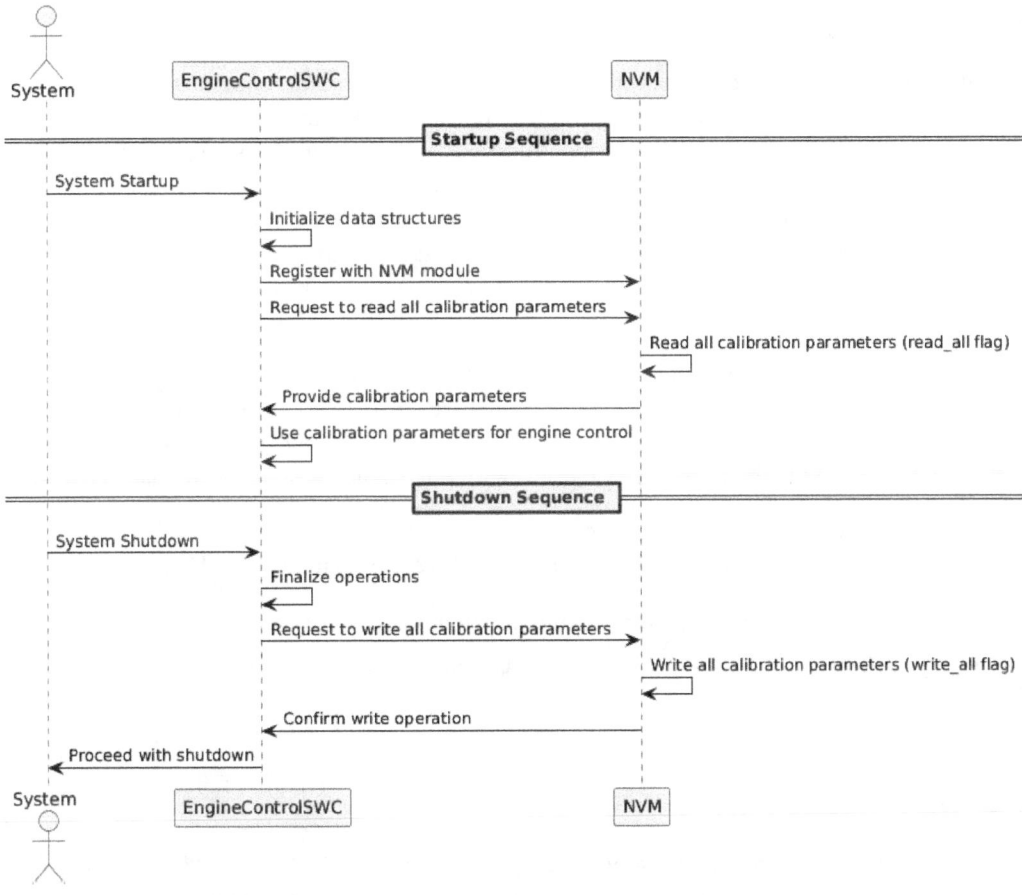

Figure 9.4 – Abstract sequence diagram for NVM usage in a system

Explanation: The selection of the `read_all` flag ensures that all calibration parameters are read during startup, guaranteeing that the ECU has access to the latest configuration.

Similarly, choosing the `write_all` flag ensures that all changes made during runtime are written back to NVM during shutdown, maintaining data integrity across power cycles.

By utilizing NVM in this manner, `EngineControlSWC` ensures that calibration parameters remain consistent and reliable, contributing to optimal engine performance and emissions control.

Now that we have understood the NVM stack and the components of NVM, it is essential to explore the mode manager for BSW. Understanding the BSWM is crucial as it plays a vital role in managing the various modes of operation within the automotive software architecture.

Mode management

The **BSWM** in AUTOSAR handles mode management by processing mode requests from both BSW modules and application layers according to configured rules. It performs specific actions based on these rules. It coordinates the operational modes of various BSW modules and application layers, ensuring seamless communication and functionality across the system. For example, the BSWM manages ECU mode changes, activates/deactivates data units in the **PDU router** (**PDUR**) module, and coordinates with the **Diagnostic Communication Manager** (**DCM**) for mode switches. During startup and shutdown, the BSWM directs the NVM to read/write necessary data blocks, ensuring proper data handling and system functionality.

The BSWM is highly configurable, allowing it to be tailored to specific application needs. It operates based on a set of rules defined during the configuration phase. These rules determine how the BSWM responds to mode requests and which actions to execute. The flexibility in configuration makes the BSWM adaptable to various automotive applications and ensures it can handle different operational scenarios.

As a mode manager, the BSWM provides a framework mechanism. The specific functional logic mainly depends on the user's configuration implementation. The BSWM is mainly divided into two parts:

- **Mode arbitration**: This is used to determine which mode it is in
- **Mode control**: This is used to perform corresponding actions

Let's learn more about them.

Role and functionality

The BSWM serves as an arbitrator for mode requests from different SWCs and BSW modules. It processes these requests based on pre-defined rules and executes corresponding actions. These actions include activating or deactivating communication protocols, managing diagnostic states, and controlling the state of the ECU.

The BSWM functionality is split into two different tasks:

- **Mode arbitration**: This task involves receiving mode indications from SWCs or BSW modules and evaluating if a mode switch can be performed based on rule arbitration. The rules, which are simple Boolean expressions, determine if a switch can occur with minimal runtime impact. The results from these evaluations are stored internally to know which action lists to execute.
- **Mode control**: This task executes mode switches by performing a set of actions, known as **action lists**, which include mode-switching operations for other BSW modules.

Figure 9.5 illustrates the process of mode request handling and control within the BSWM in the AUTOSAR framework. It shows how a mode request from an SWC is processed through the **runtime environment** (**RTE**), evaluated in the Mode Arbitration block, and executed through the Mode Control block.

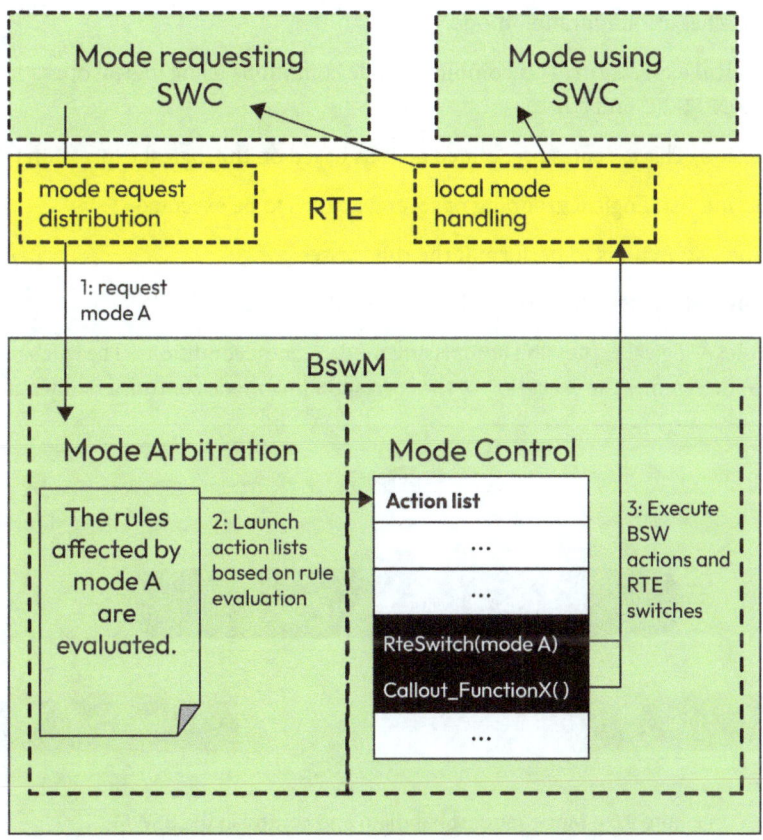

Figure 9.5 – Mode request handling and control in the BSWM

194 Dealing with Memory and Mode Management

Here is how the process goes:

1. **Mode request**: An SWC initiates a mode switch request (e.g., mode A) through the RTE.
2. **Mode arbitration**: The BSWM evaluates the mode switch request against predefined rules. These rules determine if the requested mode change can occur.
3. **Action execution**: Based on the evaluation, the BSWM triggers the corresponding action list in mode control, such as `RteSwitch(mode A)` or `RteSwitch(mode X)`.
4. **Notification**: The mode change is communicated to the relevant SWC that utilizes the requested mode

The typical implementation of the BSWM and the configuration involved are as follows:

- **BSW mode condition**: This checks if the mode request or indication from the SWC or BSW module matches a configuration mode.
- **BSWM logical expression**: This combines mode conditions using logical operations, such as `AND`, `OR`, `XOR`, `NOT`, and `NAND`.
- **BSWM action**: This executes predefined actions based on the logical expression's result.
- **BSWM action list**: Logical grouping of several actions to be executed.
- **True action list**: Actions to perform if the rule passes.
- **False action list**: Actions to perform if the rule fails.
- **BSWM rule**: A logical expression comprising mode request conditions. The rule's result (`True` or `False`) determines the execution of the corresponding mode control, as shown in *Figure 9.6*:

Figure 9.6 – Mode request handling and control in the BSWM

In BSWM mode, conditions are evaluated like an `if` statement. The mode condition checks if a request matches a configured mode. Logical expressions combine these conditions (e.g., using AND or OR) to form rules. If the rule evaluates to true, specific actions (such as function calls) are executed from the TRUE action list; if false, actions from the FALSE action list are executed. This structured process ensures efficient and accurate mode management. *Figure 9.7* illustrates an arbitration rule within the BSWM of the AUTOSAR framework, highlighting the decision-making process based on logical conditions.

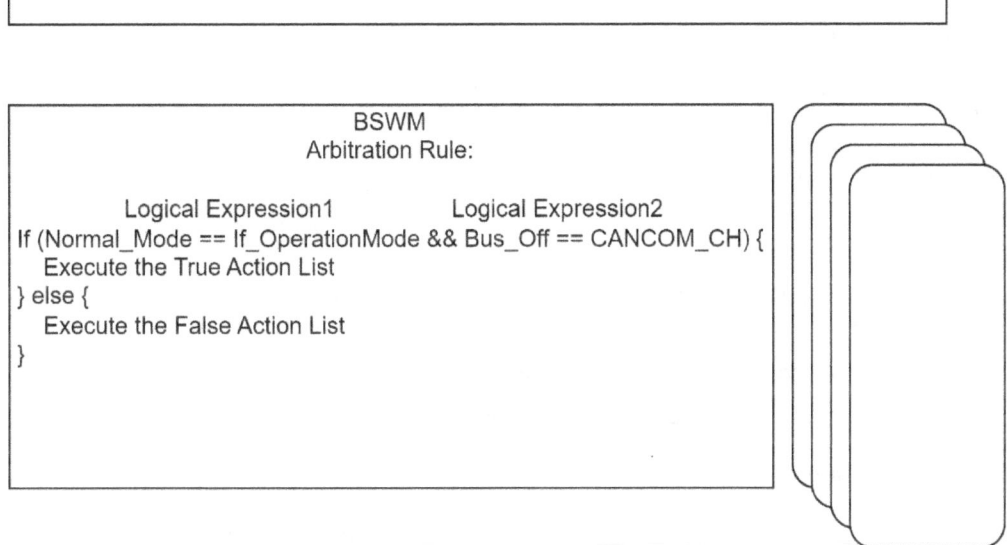

Figure 9.7 – Mode request handling and control in the BSWM

The BSWM uses these rules to manage the mode and state transitions of BSW modules. The arbitration rule evaluates two logical expressions: whether the operation mode is in normal mode and whether the evaluation for the **Controlled Area Network** (**CAN**) channel is `Bus_Off`. Depending on the evaluation, the BSWM executes either the True action list if both conditions are met or the False action list if they are not. This mechanism ensures appropriate actions are taken based on the current system state, maintaining system reliability and correct operation. Now, let's jump into a use case and learn about how the BSWM fits into a system.

> **Note**
>
> The BSWM evaluates modes and considers inputs from other modules. Modules such as **CAN State Manager (CANSM)**, **Communication Manager (COMM)**, and **NVM Manager (NvmM)** report their internal modes to the BSWM example communication channel state. These BSW module modes can be utilized to create different rules based on various modes, either from BSW modules or from the application SWCs. This approach allows for flexible and robust arbitration and decision-making processes within the automotive ECU.

Use case for the BSWM – Vehicle shutdown process

Scenario: Managing the shutdown process of an automotive ECU.

Objective: Ensuring a smooth and safe shutdown sequence, preserving critical data, and transitioning all modules to a safe state.

Here is the step-by-step use case:

1. **Mode request**:

    I. An application SWC has decided that the ECU must shut down when the driver turns off the ignition.

    II. The SWC requests Shutdown via the RTE switch port.

    III. This mode request is sent to BSWM for arbitration.

2. **Mode arbitration**:

    I. The BSWM evaluates the request based on predefined rules, such as ensuring the vehicle is stationary.

    II. If conditions are met, the BSWM proceeds with the shutdown transition.

3. **Action execution**:

    I. The BSWM triggers the corresponding action list for the Shutdown mode, once conditions are evaluated to be true.

    II. This includes instructing the NvmM to save any critical data to EEPROM or flash memory to preserve it across power cycles.

4. **Coordination with other modules**:

    I. The BSWM coordinates with the **Communication Manager (ComM)** to safely deactivate communication channels.

    II. It also interacts with the **ECU State Manager (ECUM)** to transition the ECU to a low-power state (If required).

5. **SWC notification**:

    I. The RTE notifies relevant SWCs about the mode switch.

    II. SWCs perform necessary pre-shutdown tasks, such as saving user settings or stopping ongoing processes.

6. **Finalization**:

    I. The BSWM ensures that all modules confirm the successful completion of their shutdown tasks.

    II. The ECU can then transition to a complete shutdown state.

Figure 9.8 shows the sequence diagram for the preceding use case:

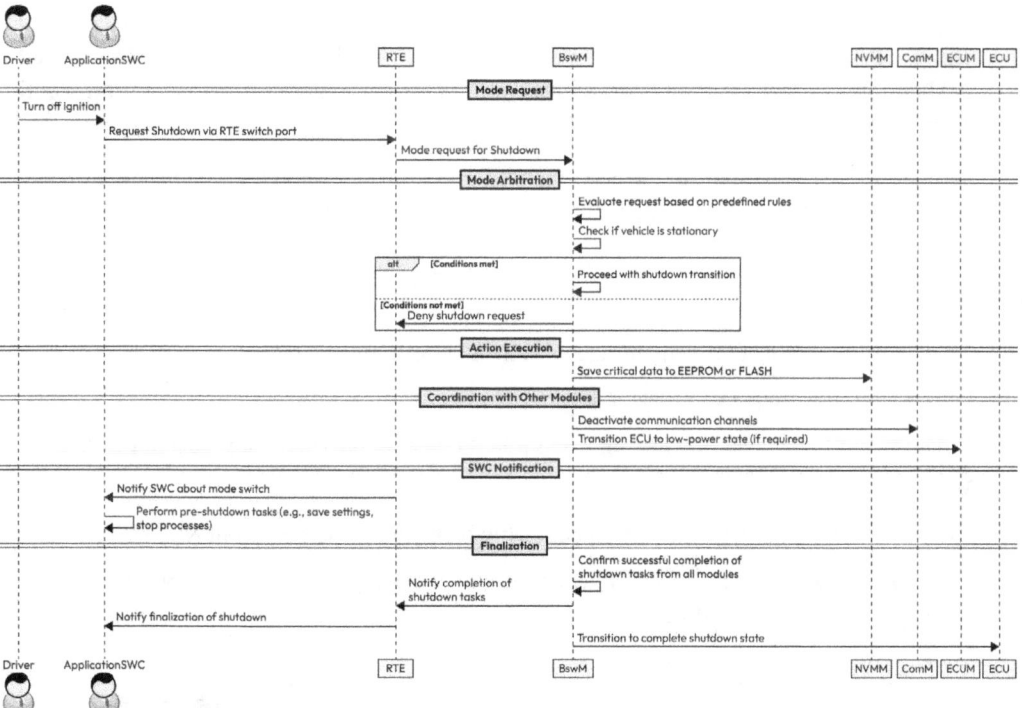

Figure 9.8 – Sequence diagram for a safe ECU shutdown

Ensuring a safe and orderly shutdown process, protecting critical data through NVM management, and enhancing overall system reliability by coordinating the shutdown sequence across all modules is helpful as it prevents data loss, maintains system integrity, and ensures that all components transition smoothly to a safe state, ultimately improving the vehicle's reliability and user experience.

Summary

This chapter on the NVM stack within the AUTOSAR framework provided developers with a comprehensive understanding of how NVM is managed in automotive software development. It highlighted the critical role of NVM in ensuring the persistence and integrity of vital data, configurations, and diagnostic information across power cycles and system resets.

By delving into the architecture and functionalities of the NVM stack, you gained insights into the mechanisms by which automotive SWCs interact with NVM. The chapter elucidated the process of data storage and retrieval through the NvmM module, emphasizing the importance of efficient storage mechanisms and error-handling strategies.

The discussion on basic storage objects, block management types, and common flags offered you practical insights into configuring NVM for diverse automotive applications. By illustrating use cases and providing examples, the chapter enabled you to grasp the significance of NVM in ensuring data integrity, reliability, and fault tolerance in automotive ECUs. Additionally, this chapter covered the BSWM and its rule-based mode management in BSW, explaining how it manages modes and coordinates actions.

The next chapter will discuss some of the best practices, other use cases, and hints on how to navigate further into AUTOSAR, serving as the concluding chapter of the book.

Questions

1. What is the purpose of implementing NVM in an automotive ECU?
2. How does a SWC request for data storage?
3. What are the advantages of selecting a redundant block management approach for NVM?
4. What is an NVM block?
5. How does the `read_all` flag influence startup behavior in the context of NVM?
6. How does the `write_all` flag influence shutdown behavior in the context of NVM? What is the role of a BSWM module?

Get This Book's PDF Version and Exclusive Extras

Scan the QR code (or go to `packtpub.com/unlock`). Search for this book by name, confirm the edition, and then follow the steps on the page.

Note: Keep your invoice handy. Purchases made directly from Packt don't require one.

10
Wrapping Up and Extending Knowledge with a Use Case

In the intricate ecosystem of AUTOSAR, where reliability, performance, and safety are paramount, mastering the development and configuration of an automotive **Electronic Control Unit** (**ECU**) reveals the true power of standardized software architecture. This chapter aims to bring together the essential concepts covered throughout this book, extending your knowledge by exploring areas such as the diagnostics stack and the time synchronization stack—stacks that are invaluable but have previously been unexamined.

Beyond revisiting these advanced topics, this chapter will empower you with practical guidance on navigating and interpreting AUTOSAR documentation, a useful skill for any developer seeking to deepen their understanding and stay at the forefront of this evolving standard. Moreover, we'll discuss the next steps you should consider as you progress in your AUTOSAR journey.

To anchor these concepts to a tangible context, we'll explore an abstract use case: the design of an auto-parking system. This example will illustrate how the diagnostics and time synchronization stacks, along with other components of the AUTOSAR architecture, can be applied to a real-world project, demonstrating the seamless integration of these layers. Although this use case has intentionally been summarized to serve as a focused illustration, it effectively bridges the gap between theoretical knowledge and practical application. By the end of this chapter, you'll have a clearer vision of how to continue your AUTOSAR journey, equipped with both the foundational knowledge and the practical insights necessary for future challenges.

In this chapter, we will discuss the following:

- Overview of diagnostics
- Overview of time synchronization
- How to read the standards
- What's next?
- Case study – development of an autonomous parking assist system

Overview of diagnostics

In this book, we've embarked on a journey through the foundational aspects of AUTOSAR, beginning with its history and motivations, followed by an in-depth look at the **runtime environment** (**RTE**), **software component** (**SWC**), and core modules within the **basic software** (**BSW**) layer, such as communication, security, and memory management. These elements form the backbone of AUTOSAR and are key to understanding how this architecture standardizes and enhances automotive software development.

However, it's important to recognize that there are other essential components within the AUTOSAR framework that we have not covered in detail. Notably, diagnostics are critical to automotive systems and vital for maintaining vehicle safety and performance. While we haven't deeply examined diagnostics and other advanced modules, it's important to note their significance.

Diagnostics in automotive systems are key to ensuring the reliability, safety, and performance of vehicles. The primary goals of diagnostics are to detect faults, report them accurately, and provide mechanisms for repairing and maintaining the vehicle ECUs. Effective diagnostic systems help with the following:

- **Early fault detection**: Identifying issues early on prevents minor problems from escalating into major failures, thereby improving vehicle safety and reliability.
- **Maintenance and repair**: Diagnostics facilitates efficient maintenance and repair processes by providing precise fault information and reducing downtime and associated costs.
- **Compliance with regulations**: Many regions have strict regulations regarding vehicle emissions and safety. Diagnostics help ensure compliance with these standards.
- **Customer satisfaction**: By ensuring vehicles operate smoothly and safely, diagnostics enhance customer satisfaction and trust in the brand.

As we have grasped the importance of diagnostics in a vehicle, let's discuss the modules that compose the diagnostics stack, starting with the **Diagnostic Communication Manager** (**DCM**) module.

Diagnostic Communication Manager (DCM) in detail

DCM is a central component in the AUTOSAR diagnostics stack. Its primary role is to ensure that the vehicle's internal systems—such as the engine control unit, transmission control, and other essential electronic units—can communicate effectively with external diagnostic tools. This allows technicians to accurately diagnose issues, perform maintenance, and update software, ensuring the vehicle runs smoothly and safely. Key features of DCM include the following:

- **Diagnostic session management**: DCM handles different diagnostic sessions, such as default, programming, and extended sessions, each offering varying access levels to diagnostic services. It ensures that the vehicle is in the correct session for the requested diagnostic operation, controlling transitions between these sessions as needed.

> Extended session
>
> The primary purpose of the extended session is to provide access to features that typically require higher security or are more intrusive to the vehicle's normal operation. Examples of functions that might be accessible only in an extended session are the motor calibration and tuning adjusting parameters in the ECU, such as those for engine tuning or motor control.

- **Security and access control**: To protect against unauthorized access, DCM manages security access levels and ensures that sensitive diagnostic operations are performed only after the appropriate security mechanisms, such as seed/key algorithms, are satisfied.
- **Processing diagnostic requests**: DCM interprets diagnostic requests from external tools and forwards them to the appropriate internal modules.
- **Service handling**: This supports a wide range of diagnostic services as defined by standards such as **Unified Diagnostic Services** (**UDS**) and J1939.

> Note on UDS
>
> UDS is a standard protocol (ISO 14229) used in automotive and other industries for diagnosing and configuring vehicles or devices over a communication network, typically a **controller area network** (**CAN**). UDS defines a set of diagnostic services that enable functions such as reading **diagnostic trouble codes** (**DTCs**), monitoring the status of vehicle components, performing software updates, and configuring parameters. The protocol standardizes communication between a diagnostic tool and the vehicle's ECUs, facilitating consistent and reliable diagnostics across different vehicle makes and models.

Figure 10.1 shows the process of a diagnostic session where a tester tool sends a diagnostic request via the CAN. The request is routed through various layers, including the CAN, **CAN Transport Protocol (CanTp)**, and **Protocol Data Unit Router (PDUR)**, before reaching DCM. DCM then interacts with the SWC via the RTE to request a memory read operation. The response follows a similar path back to the tester tool, ensuring the secure and reliable communication of diagnostic data between the vehicle's internal systems and the external diagnostic tool.

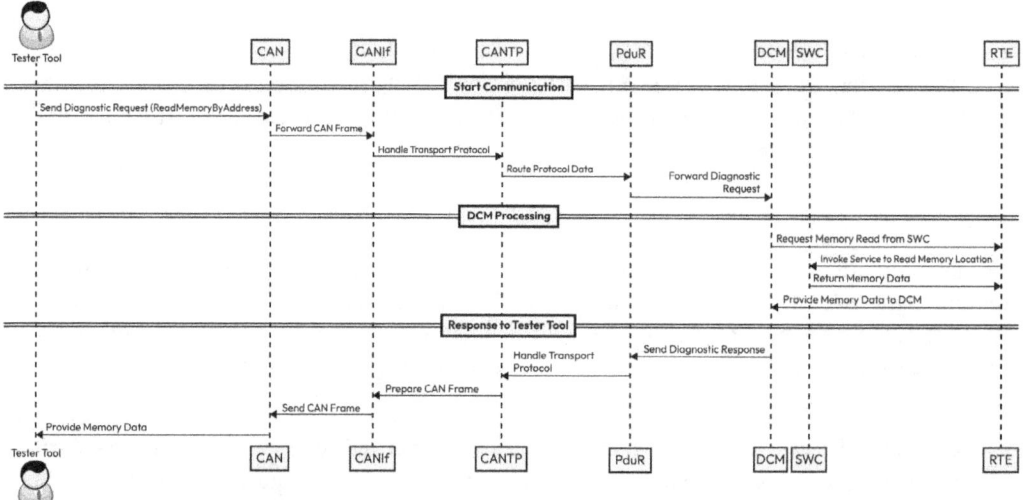

Figure 10.1 – Sequence diagram illustrating DCM in an AUTOSAR architecture

After understanding DCM's role in managing and translating communication, let's now explore **Diagnostic Event Manager (DEM)**, which completes the picture by gathering, debouncing, and managing diagnostic events within the vehicle's ECUs.

Diagnostic Event Manager

DEM is part of the diagnostics stack. It handles the management of diagnostic events and the storage of fault information. In an AUTOSAR system, while DCM oversees the communication of diagnostic data, DEM is responsible for internally collecting, processing, and managing these events. It acts like the system's internal detective, continuously monitoring through another SWC, which acts as a monitor for events, recording and analyzing the vehicle's operational state to identify faults and anomalies.

Its main responsibilities include the following:

- **Fault detection**: An SWC monitors various signals and parameters to detect faults in the vehicle systems and reports these events to DEM, which then performs debouncing and other processing steps to qualify the faults.

- **Event status management**: DEM keeps track of the status of each event, including whether an event is currently active, has been confirmed, or is pending. This helps with understanding the severity and status of the detected faults.

- **DTC management**: It records DTCs whenever a fault is detected, which is essential for diagnosing issues.

> **A note on DTCs**
>
> A DTC is a standardized code read by a vehicle's **onboard diagnostics** (**OBD**) system to indicate specific faults or issues within the vehicle's electronic systems. Each DTC consists of a combination of letters and numbers that correspond to problems, such as engine misfires, sensor failures, or communication errors. For instance, the code `P0301` indicates a misfire in cylinder 1. In practice, a software monitor within the ECU continuously checks for such errors and, upon detecting a fault, records it as a DTC. These DTCs are then stored in the ECU's memory and can be retrieved later by diagnostic tools for analysis, allowing for precise fault identification and efficient debugging and repair processes.

- **Event memory**: DEM maintains an event memory to keep track of all diagnostic events, which can be accessed during service and maintenance.

Figure 10.2 illustrates the AUTOSAR architecture, focusing on the interaction between key components across different layers. DEM, centrally located in the BSW layer, is responsible for collecting, processing, and storing diagnostic events. For example, a software monitor in the ECU might detect a fault and log it as a DTC in DEM. Later, DCM can retrieve this DTC, allowing external diagnostic tools to read and analyze the stored fault information.

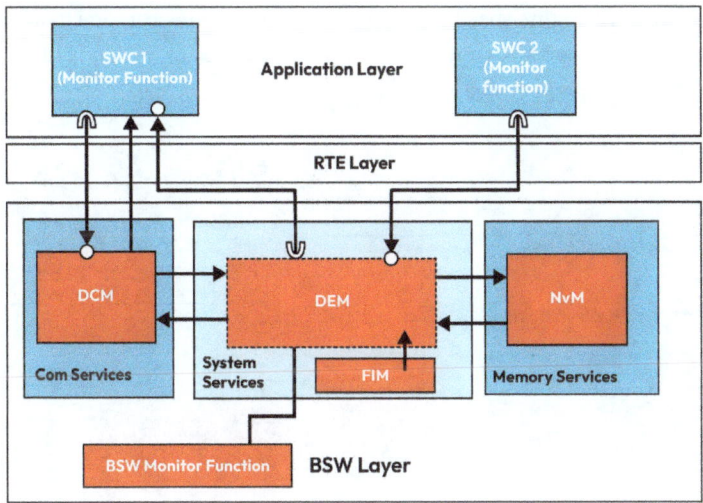

Figure 10.2 – Interaction between DCM and DEM within the AUTOSAR architecture

Having explored DEM and DCM, which focus on production-level diagnostics, it's important to also consider internal diagnostics within the system. This is where the **Default Error Tracer (DET)** comes into play. The DET operates at another level of diagnostics. It provides critical functionality for detecting and tracing errors during both development and runtime.

Default Error Tracer (DET)

The DET is a diagnostic method used during the development phase of automotive software. It helps developers identify and trace errors in the system, guaranteeing the robustness and reliability of the final product. Key responsibilities include the following:

- **Error detection**: Monitors the system for development errors
- **Error logging**: Records detailed information about detected errors
- **Debugging support**: Provides valuable insights for developers to debug and fix issues effectively
- **Internal error detection**: Continuously monitors the software's internal operations, detecting issues such as incorrect function calls, parameter mismatches, or failed assertions

The DET's focus is on the software development life cycle rather than on the vehicle's operational diagnostics. It serves as a critical tool for developers by ensuring that the software behaves as expected before it is deployed in a car. As the AUTOSAR standard puts it: "*The Default Error Tracer provides immediate detection and tracing of errors, facilitating robust software development and reducing the risk of undetected faults during runtime.*"

Imagine you're developing a software module for an ECU in a vehicle. During the development phase, one of the functions, let's say `CalculateEngineTorque()`, is supposed to be called with a valid range of parameters, such as engine speed and throttle position. However, due to a bug in the code, this function is sometimes called with a negative engine speed value, which is invalid. *Figure 10.3* shows such a use case as a sequence diagram.

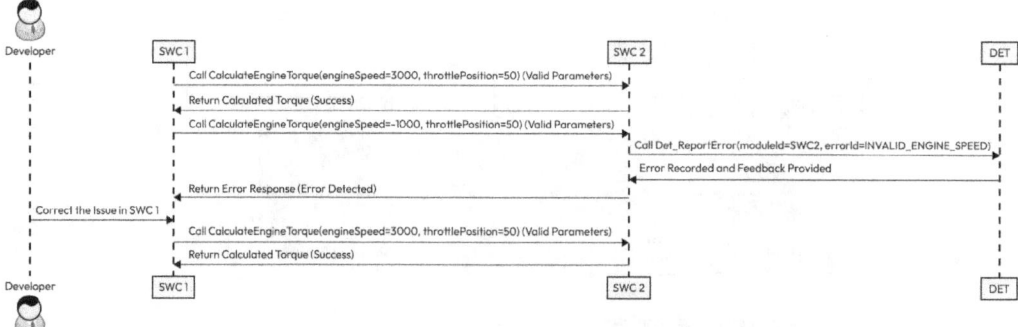

Figure 10.3 – Sequence diagram demonstrating the DET functionality

Failure Management

The **Failure Management (FiM)** module interacts with DEM to manage the status of diagnostic events and control functions based on the fault conditions. FiM is designed to enable or disable certain functions based on the occurrence of specific faults, helping to maintain safety and vehicle performance.

The use case in *Figure 10.4* illustrates this process, showing how FiM evaluates conditions and controls function permissions dynamically in response to detected faults. Based on the fault information retrieved, FiM grants or denies permission to these functions.

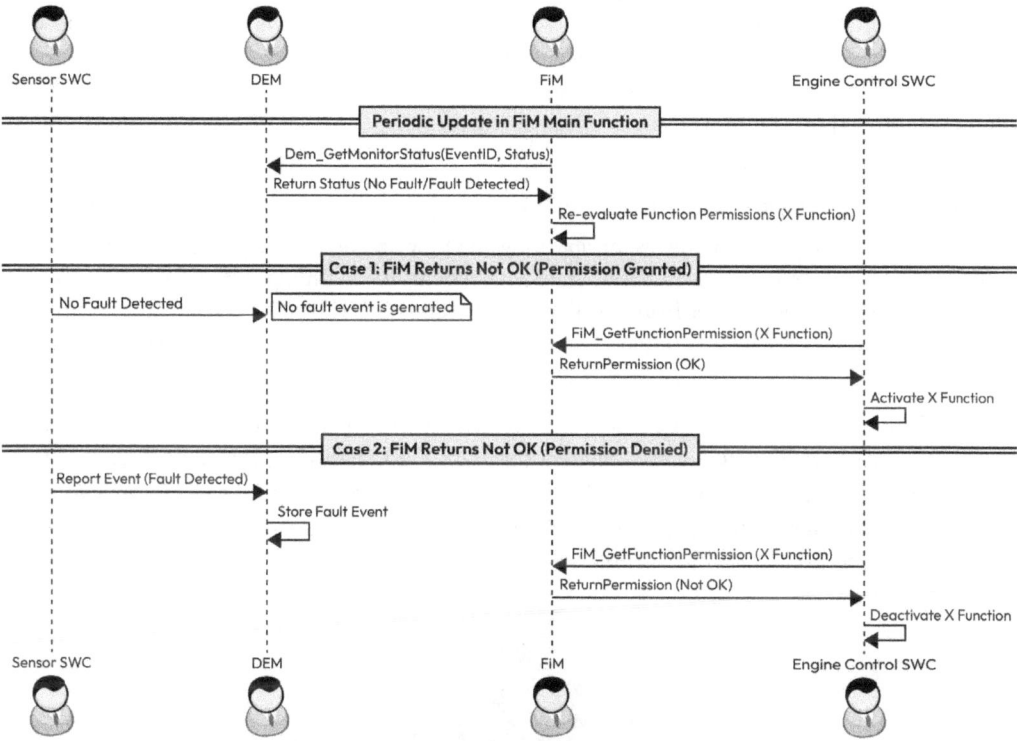

Figure 10.4 – Sequence diagram demonstrating the DET functionality

With the overview of diagnostics—covering DCM, DEM, DET, and FiM—complete, it's important to recognize that we've only scratched the surface. Each module offers extensive features and configurations to achieve robust diagnostics in automotive systems. Now, let's transition to another vital aspect: time synchronization. In systems with multiple ECUs, precise time alignment is essential for coordinated operation and communication. Next, we'll explore how time synchronization can be effectively implemented in an automotive context.

Overview of time synchronization

Time synchronization in AUTOSAR is a set of components designed to ensure accurate and consistent timekeeping across different ECUs within a vehicle. This is achieved through dedicated time synchronization protocols and mechanisms, providing a common time base for various applications and systems to enable precise coordination and timing-dependent operations. In automotive systems, time-critical applications are often distributed across multiple ECUs connected via different in-vehicle networks such as CAN, Ethernet, and FlexRay, each operating with its own source clock. Synchronizing these ECUs is essential to achieve time-sensitive functionalities. The challenge in a mixed network architecture is synchronizing nodes across different protocols connected via a gateway. AUTOSAR addresses this challenge with the concept of a Time Gateway, which ensures synchronization across diverse networks, enabling seamless time-sensitive operations, and this is where the AUTOSAR modules help to achieve this goal

> **What are the Time Master and Time Slave?**
>
> The **Time Master** is a central node responsible for generating and distributing the reference time across the network, ensuring all nodes operate with synchronized timing. The **Time Slave** is a node that receives time updates from the Time Master and adjusts its own clock to align with the reference time, allowing coordinated and time-consistent operations across the network.

Figure 10.5 depicts the AUTOSAR time synchronization framework, highlighting the four key modules responsible for synchronizing time across different communication buses: EthTSyn for Ethernet, **CAN Time Synchronization (CanTSyn)** for CAN, and FrTSyn for FlexRay. Central to this framework is **Synchronized Time-Base Manager (StbM)**, which coordinates these modules to ensure a consistent and synchronized time base across the entire system.

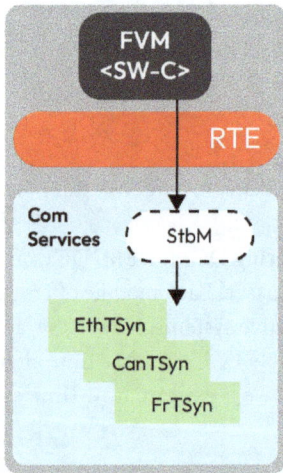

Figure 10.5 – Time synchronization architecture

Imagine an orchestra where each musician represents an ECU. In an orchestra, the conductor ensures that all musicians play in sync, following the same tempo and rhythm. If the conductor failed to synchronize the musicians, the resulting performance would be chaotic, with instruments playing out of tune and time with each other.

Similarly, in a vehicle, the **sync module** acts like the conductor, ensuring that all ECUs are synchronized to a common time base. Just as a conductor provides cues to keep the musicians in sync, sync modules provide accurate timestamps and time alignment across the network nodes.

Accurate time synchronization is essential for several reasons:

- **Event correlation**: Precise timestamping allows for the accurate correlation of events across different ECUs, aiding in fault diagnosis and system analysis
- **Real-time systems**: Many vehicle systems, such as **Advanced Driver Assistance Systems (ADASs)** and autonomous driving, rely on real-time processing and require precise timing for correct operation
- **Sensor fusion**: Combining data from multiple sensors often involves timestamping and synchronization to ensure accurate results
- **Communication protocols**: Some communication protocols, such as Ethernet, require time synchronization for efficient and reliable data transfer
- **Logging and recording**: Precise timestamps are vital when logging and recording data for accurate analysis and replay

Synchronized Time-Base Manager (StbM) module

The StbM module is a fundamental component within the AUTOSAR architecture responsible for managing time synchronization within a vehicle. It provides the following core functionalities:

- **Time-based management**: Maintains a local time base on each ECU, ensuring accurate timekeeping.
- **Synchronization of runnable entities**: Runnables can be configured to execute at specific times based on the synchronized time base provided by StbM. By doing so, the execution of these runnables can be precisely coordinated across various ECUs.
- **Time domain management**: Supports multiple time domains within a vehicle, allowing for different time references based on specific system requirements.
- **Time offset calculation**: Determines the offset between the local time base and the synchronized time.
- **Time adjustment**: Adjusts the local clock based on calculated offsets to maintain synchronization.
- **Time information provision**: Provides time information to other SWCs as required.

One of the use cases I always like to describe is how to achieve synchronization between ECUs using StbM and why it is sometimes needed in special cases.

Use case – synchronizing radar ECUs with StbM

Consider a scenario where a vehicle has multiple radar ECUs, each responsible for emitting **radio frequency** (**RF**) signals to detect objects in the vehicle's surroundings. To avoid interference between the radar signals, it is required that each radar ECU triggers its RF signal with a precise time offset of 40 microseconds (µs) from the others.

Figure 10.6 illustrates a straightforward interaction where two radar ECUs synchronize with the Time Master provider ECU, schedule their tasks with the necessary time offsets, and execute these tasks independently.

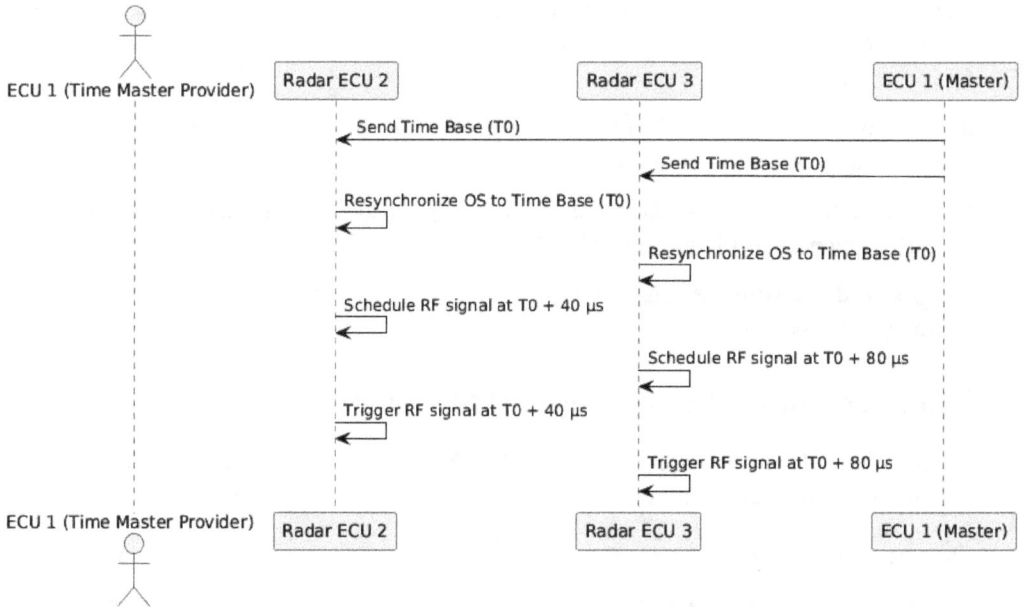

Figure 10.6 – Time synchronization architecture

You have been introduced to StbM and understand its role as the central timekeeper that interacts with the OS to reschedule tasks or adjust the OS schedule table. Let's now explore CanTSyn. This module is specifically responsible for CAN bus time synchronization and obtaining the necessary time data. Similar modules, such as EthTSyn and FrTSyn, perform comparable functions, with variations tailored to the specific communication bus they support.

CAN Time Synchronization (CanTSyn)

The CanTSyn module is responsible for handling the distribution and synchronization of time information over CAN buses. However, simply transmitting the time information from the Time Master to the Time Slaves through a broadcast CAN message can lead to inaccuracies due to CAN-specific factors such as message arbitration and delays within the BSW.

To address these challenges, the CanTSyn module uses PTP synchronization, employing a process to ensure precise timing across the network. *Figure 10.7* shows how this process is achieved.

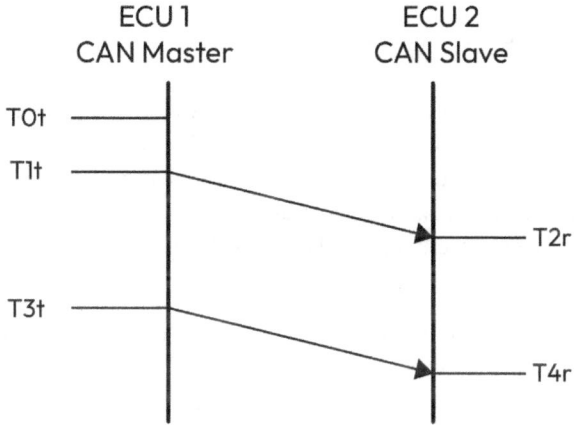

Figure 10.7 – CanTSyn mechanism

The preceding figure highlights a four-step process, which is described as follows:

1. **SYNC message**: The Time Master ECU broadcasts a SYNC message with the current ECU time (**T0t**). It records the exact transmission time (**T1t**) the message was transmitted by the driver.
2. **Reception by Time Slaves**: Time Slaves receive the SYNC message and a timestamp of the exact reception time (**T2r**).
3. **Follow-Up (FUP) message**: The Time Master then sends an FUP message containing the time difference between the actual transmission and the time indicated in the SYNC message (**T1t–T0t**).
4. **Time adjustment**: Slaves combine the information from the SYNC and FUP messages to accurately adjust their local clocks, aligning precisely with the Time Master, and provide this as an indication to StbM to keep the time.

In the previous sections and chapters, we covered the various modules within the AUTOSAR stack, providing an overview of their roles and interactions. One challenge I've encountered in the industry is that while many questions and issues are addressed in the AUTOSAR standards, some professionals might overlook or miss the importance of thoroughly reading and understanding these core documents. Mastering the standard and avoiding common pitfalls is impossible without fully grasping the art of reading documentation and knowing where to search for information or hints that you might be missing.

With this in mind, let's move on to the next section, where we'll explore the core AUTOSAR documents, providing guidance on how to access, interpret, and apply these essential resources in our work.

How to read the standards

Regardless of which configurator you choose, you'll quickly realize that implementing requirements goes beyond just using tools—it's about deeply understanding the specifications that define the AUTOSAR standard. These specifications serve as a common blueprint, outlining how each module should behave and the configuration parameters that allow you to extend and customize a module's functionality. The tools are like instruments used by builders to modify and adapt these blueprints, enabling you to tailor the modules to your specific needs.

For example, in the case of the COM module, the specifications define the configuration parameters that could influence the transmission of a PDU. This is helpful for developers and architects to configure the COM module and manage communication signals efficiently and properly. The configurator tools allow you to adjust parameters such as signal timeouts or data packaging options, but these tools are only as effective as your understanding of the underlying requirements.

Additionally, these configurations can be shared in ARXML files, much like blueprints that different builders can use to ensure their work aligns perfectly with the overall design. This allows other modules to read these settings and automatically adapt their internal configurations, providing seamless integration. This is a key reason why different vendors can collaborate so effectively—they all work from the same blueprint, with a shared understanding of the specifications and a consistent approach to module configurability.

Building on what was mentioned earlier, the following section will guide you through some of the common AUTOSAR documents and their contents, helping you understand the specifications and navigate them effectively.

Key AUTOSAR documents

This book's main goal has been to give you a solid introduction to the fundamental concepts and most common stacks in AUTOSAR. To gain a comprehensive understanding and explore areas not covered here, I encourage you to refer to the detailed AUTOSAR specifications available at `https://autosar.org`. These documents will help you get a clearer understanding of advanced topics and additional modules that haven't been discussed in this book:

- **Software Specification (SWS)**:

  Purpose: Defines the software module's functionality

  Contents:

  - **Overview**: General introduction and functional overview of the module.
  - **Requirements**: Functional specification
  - **Interfaces**: Detailed descriptions of APIs, including input and output parameters.
  - **Behavior**: Detailed description of the module's behavior, state machines, sequence diagrams, and flow diagrams.
  - **Configuration parameters**: List, as well as explanations, of configuration options available.
  - **Example**: An example that you can refer to is **Communication Manager (ComM)**, stored at `https://www.autosar.org/fileadmin/standards/R23-11/CP/AUTOSAR_CP_SWS_COMManager.pdf`. The document shows an SWS for ComM in AUTOSAR. Its purpose is to define the functionality, requirements, and interfaces of the ComM module, which manages the communication modes for the communication buses within an ECU. This document provides detailed guidelines on how ComM interacts with other modules, handles communication mode transitions, and ensures efficient data exchange within the vehicle's network.

> **Note**
>
> For each AUTOSAR module we've discussed, you can refer to its corresponding SWS document. These documents offer a comprehensive understanding of the module, including detailed descriptions of interfaces, state machines, sequence diagrams, and its interactions with other modules. They are essential for gaining a deeper insight into how each module operates within the AUTOSAR framework.

- **Template Protocol Specification (TPS):**

  Purpose: Define the format, structure, and content for specific templates used within the architecture

  Contents:

  - **Introduction:** General information about the document's purpose.
  - **Guidelines:** Detailed implementation guidelines.
  - **Example:** The document `https://www.autosar.org/fileadmin/standards/R23-11/CP/AUTOSAR_CP_TPS_SoftwareComponentTemplate.pdf` is a **Technical Protocol Specification** (**TPS**) for the SWC template in AUTOSAR. Its purpose is to define the structure and content of SWCs within the AUTOSAR architecture, providing a standardized template that ensures consistency across different implementations. This document outlines how SWCs should be described, including their interfaces, attributes, and relationships, to support seamless integration and interoperability within AUTOSAR-compliant systems.

- **Requirement Specification (RS):**

  Purpose: Outline the high-level requirements that the standard or a specific component or a stack must fulfill:

  - **Requirement IDs:** Unique identifiers for each requirement
  - **Description:** Detailed explanation of each requirement
  - **Priority:** Importance level of each requirement
  - **Dependencies:** Any dependencies or constraints related to the requirement

- **Explanation (EXP):**

  Purpose: Provides detailed insights and justifications for specific concepts, design decisions, or processes within the AUTOSAR framework. These documents are intended to help developers and engineers understand the reasoning behind certain standards, offering context that aids in the correct implementation and integration of AUTOSAR components.

  Example: `https://www.autosar.org/fileadmin/standards/R23-11/CP/AUTOSAR_CP_EXP_FirmwareOverTheAir.pdf` provides an in-depth overview of the **Firmware Over-the-Air** (**FOTA**) update process within the AUTOSAR framework. It explains the concepts, architecture, and processes involved in remotely updating firmware in vehicles, including security measures, communication protocols, and system requirements.

Navigating AUTOSAR documentation can be a daunting task due to its comprehensive and detailed nature. To make this process more manageable, it is important to adopt a structured approach. This section provides a guide on how to effectively read and understand AUTOSAR documents, ensuring that you grasp the essential concepts and detailed specifications needed for your work. By starting with the basics, focusing on relevant sections, and reading methodically, you can efficiently navigate through the extensive AUTOSAR documentation and apply its principles to your projects:

- Start with the basics:
  - **Overview documents**: Begin with the introductory and overview documents that provide a high-level understanding of AUTOSAR and its architecture, goals, and core principles. These documents set the stage for more detailed specifications.
  - **Glossary**: Familiarize yourself with the terminology used in AUTOSAR specifications. Understanding the specific terms and abbreviations is important for comprehending detailed documents.
- Focus on relevant sections:
  - **Identify key areas**: Depending on your role and the specific aspects of AUTOSAR you are dealing with, focus on the relevant sections.
  - **Use cases**: Look for use cases and examples that illustrate how the specifications are applied in real-world scenarios. This can provide context and make the specifications more relatable.
- Read methodically:
  - **Top-down approach**: Start with high-level descriptions and gradually move to detailed specifications. This helps in building a comprehensive understanding without getting overwhelmed.
  - **Reference models**: Keep referring to architecture diagrams and reference models provided in the documents. Visual representations can aid in understanding complex interactions and data flows.

As we conclude our comprehensive exploration of AUTOSAR, it's important to recognize that this is just the beginning of your journey through the vast and dynamic field of automotive systems.

What's next?

The world of AUTOSAR offers endless opportunities for learning, innovation, and professional growth. Here are some ways to continue your journey:

- **Stay updated with industry trends**: The automotive industry is rapidly evolving, with new advancements in autonomous driving, electric vehicles, and connectivity. Staying informed about these trends will provide valuable context and help you understand the direction in which AUTOSAR is heading. Follow industry news, attend conferences, and participate in webinars to keep your knowledge current. Monitor new releases regularly by checking the official AUTOSAR website for new releases, which are typically published twice a year.

- **Engage with the AUTOSAR community**: Join online forums, discussion groups, and professional networks focused on AUTOSAR. Engaging with the community allows you to share experiences, ask questions, and learn from the insights of others. Consider joining organizations and attending AUTOSAR consortium meetings to connect with experts and practitioners in the field.
- **Get hands-on practice**: Apply your knowledge by working on real-world projects, either through your current role or personal projects. Hands-on experience is invaluable for deepening your understanding and honing your skills. Experiment with different aspects of AUTOSAR, from developing BSW modules to implementing security mechanisms.

As we conclude, it's important to remember that while we haven't covered every module within AUTOSAR, our selective approach was designed to give you a solid foundation in the most critical aspects. This introduction should equip you with the knowledge needed to confidently embark on your journey of managing and working with AUTOSAR projects.

Let's recap what we've covered in this book with a case study on building an autonomous parking assist system. This case study will be somewhat abstract, but it will provide a clear flow—from understanding the system to defining the requirements, and even touching on the design process and identifying the key modules you might need in your ECU.

Case study – Development of an autonomous parking assist system

You have just been introduced to key aspects of AUTOSAR, including the RTE, SWCs, communication, diagnostics, security, memory management, and mode management. To solidify your understanding, let's embark on a case study that involves developing an **Autonomous Parking** (**AP**) system. This system will get a vehicle to autonomously park by integrating various sensors and control algorithms:

User story: Imagine a driver named Yahya who struggles with parallel parking in crowded urban areas. To assist Yahya, we want to develop an AP system for his vehicle. This system will detect an empty parking slot and automatically steer and control the speed of the car to park it safely, accurately, and efficiently, while providing real-time feedback to Yahya through the vehicle's **Human Machine Interface** (**HMI**).

System requirements: To build this system, we need to establish clear and comprehensive requirements that cover various aspects, including functionality, communication, diagnostics, security, memory management, and mode management.

Figure 10.8 shows an abstract view of how this system is designed.

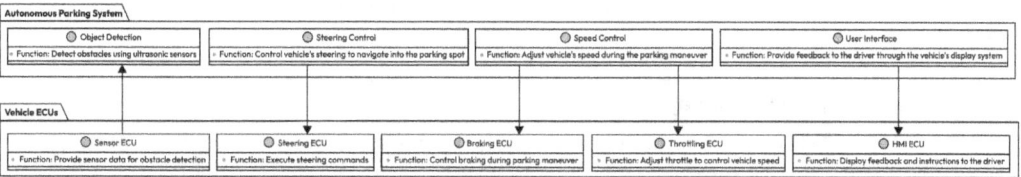

Figure 10. 8 – System diagram for the AP system

The following are the functional requirements:

- **Object detection**: The system should detect obstacles using ultrasonic sensors
- **Steering control**: The system should control the vehicle's steering to navigate into the parking spot
- **Speed control**: The system should adjust the vehicle's speed during the parking maneuver
- **User interface**: The system should provide feedback to the driver through the vehicle's display system

The following are the communication requirements:

- **Sensor data**: The system should collect data from ultrasonic sensors via CAN
- **Control commands**: The system should send steering and speed commands over CAN
- **User feedback**: The system should display status information and instructions on the dashboard

The following are the diagnostics requirements:

- **Fault detection**: The system should identify and report faults in sensors and actuators
- **DTCs**: The system should store and report DTCs for maintenance purposes

The following are the security requirements:

- **Data integrity**: The system should ensure the integrity of control commands
- **Authentication**: The system should authenticate communication between ECUs

The following memory management requirements:

- **Efficient storage**: The system should optimize memory usage for sensor data and control algorithms
- **Persistent storage**: The system should store DTCs and system logs

The following are the mode management requirements:

- **Operating modes**: The system should manage different operating modes, including active, standby, and fault

What we have discussed are all general system requirements. We can refer to the following software requirements as a starting point from which we can start defining our software and the AUTOSAR layers needed:

Requirements ID	Requirements
[AP_SW_001]	The software should process data from ultrasonic sensors to detect obstacles in real time.
[AP_SW_002]	The software should trigger appropriate responses based on obstacle detection, such as adjusting steering or stopping the vehicle.
[AP_SW_003]	The software should calculate and send steering angle commands to control the vehicle's steering during parking maneuvers.
[AP_SW_004]	The software should ensure smooth and accurate steering adjustments based on sensor data.
[AP_SW_005]	The software should manage the vehicle's speed by adjusting throttle and brake commands during the parking process.
[AP_SW_006]	The software should ensure smooth deceleration and acceleration during maneuvers.
[AP_SW_007]	The software should provide real-time feedback to the driver via the vehicle's display system, including parking status, instructions, and warnings.
[AP_SW_008]	The software should update the display based on the system's state, such as when an obstacle is detected or when parking is complete.
[AP_COM_001]	The software should implement the CAN protocol to handle communication between sensors, actuators, and control units.
[AP_COM_002]	The software should manage the transmission and reception of sensor data and control commands over CAN.
[AP_CRY_001]	The software should ensure data integrity and authenticity for control commands transmitted over the network.
[AP_NVM_001]	The software should manage non-volatile memory for storing DTCs and system logs.
[AP_NVM_001]	The software should manage non-volatile memory for storing DTCs and system logs.
[AP_GEN_001]	The ECU software should be compliant with the AUTOSAR standard and adhere to its guidelines for architecture, communication protocols, diagnostics, memory management, mode management, security, and SWC integration.

Table 10.1 – Proposed list of software requirements

The software architecture might look something like what is shown in *Figure 10.9*. To explain the rationale behind the selection of modules shown in the architecture diagram and how they satisfy the software requirements, I'll map each module to the corresponding requirements and provide a brief rationale for their inclusion in the design. This will help you understand the logical connection between the software requirements and the architectural design.

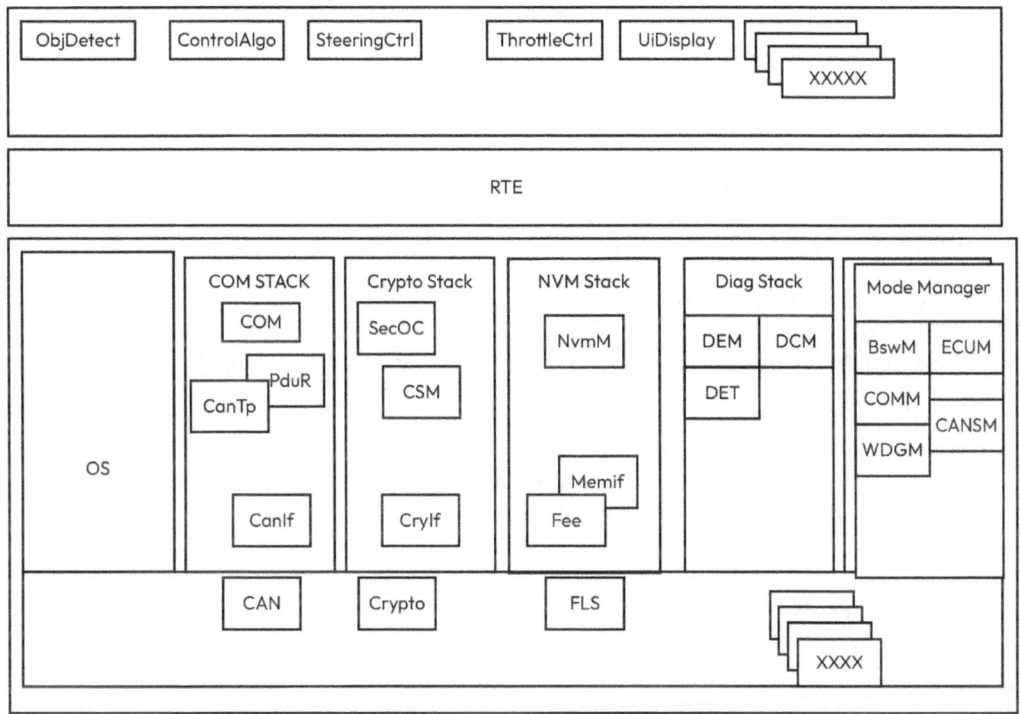

Figure 10. 9 – Proposed software architecture for the AP ECU

Let's briefly outline the design choices that have influenced this architecture.

Application layer (SWCs)

We can use the **Object Detection** (**ObjDetect**) SWC as an example of one of the software modules composing the application layer. Other SWCs cover the rest of the requirements and implement the needed interfaces to the BSW:

- **Which requirement does it cover?**: [AP_SW_001].
- **Rationale**: ObjDetect processes data from ultrasonic sensors to detect obstacles in real time. It is positioned above the RTE because it interacts directly with sensor data and influences control decisions without needing to manage hardware details, which are abstracted by the RTE.

> **Note**
> Other modules, such as **Control Algorithm (ControlAlgo)**, **Steering Control (SteeringCtrl)**, **Throttle Control (ThrottleCtrl)**, and **User Interface Display (UIDisplay)**, follow the same pattern. They handle specific vehicle control tasks or user interactions, leveraging the RTE for communication with the underlying BSW layers.

Now, let's explore some of the choices for the BSW stack.

AUTOSAR BSW modules

We are using the communication and crypto stacks as examples. Other stacks follow a similar process.

Communication stack (COM, PduR, CanTp, and CanIf)

The following requirements are covered by the communication stack:

- **Which requirements does it cover?**: [AP_COM_001] and [AP_COM_002].
- **Rationale**: The communication stack is needed for managing the exchange of data between various SWCs and the CAN, which is required for the operation of the autonomous parking system. Observe that we have selected CAN modules as we have assumed communication would take place through a CAN bus. If it were Ethernet, you might add Eth, EthIf, TcpIp, SoAd, and SD.

Crypto stack (SecOC, CSM, CryIf, and Crypto)

The following requirements are covered by the crypto stack:

- **Which requirement does it cover?**: [AP_CRY_001].
- **Rationale**: The crypto stack is needed to secure data transmitted across the communication channel.

> **Generic note**
> Similarly, the diagnostics stack (including modules such as DEM, DCM, and DET), the **non-volatile memory** (**NVM**) stack (including modules such as NvmM, MemIf, and Fee), and the mode management stack (including modules such as BswM, EcuM, ComM, WdgM, and CanSM) were selected to satisfy both the AUTOSAR requirements and the specific needs of this ECU. These stacks ensure effective fault detection, memory management, and mode transitions, which are essential for the reliability and functionality of the autonomous parking system.

Please note that the selection of modules can vary depending on the specific requirements of different ECUs. For instance, this architecture doesn't include Ethernet-related modules because they aren't needed for the current design. If Ethernet communication were required, additional modules would be integrated into the architecture to accommodate those needs. Remember, AUTOSAR architecture is flexible, and always make sure to add what you really need.

By following this mockup case study, you will gain abstract practical experience in defining requirements and designing architecture for an AP assist system using AUTOSAR principles. This exercise not only reinforces your understanding of AUTOSAR but also prepares you for real-world automotive software development challenges. Remember, this is just the beginning of your journey—continuing to study, read specifications, and gain hands-on experience is essential to mastering AUTOSAR.

Summary

Wrapping up our exploration of AUTOSAR, it's clear that this is just the beginning of our journey. AUTOSAR represents a significant advancement in automotive software engineering, promoting standardization and modularity. This book has provided an introduction to key concepts and components, including an overview of the architecture, key documents such as SWS, TPS, RS, RTE, and BSW specifications, and a detailed use case for designing and implementing a real-time control system within an automotive ECU. While the discussed fundamentals provide broad insights, they merely serve as the initial step in a much broader and more complex journey. The purpose of this book is to introduce the framework to engineers, laying the groundwork for further study and practical application. To fully harness the potential of AUTOSAR, engineers must continue to study, read the extensive specifications, and gain hands-on experience through real-world projects. As the field of automotive software continues to evolve, staying updated with industry trends and engaging with the AUTOSAR community will be essential for mastering this powerful standard.

Thank you for joining me on this journey. I hope this book has equipped you with the knowledge and confidence to navigate the complexities of AUTOSAR and apply its principles effectively in your work.

Embarking on the journey to learn AUTOSAR is a significant step toward advancing your career in the automotive industry. This book has provided you with the foundational knowledge and practical insights needed to navigate the complexities of AUTOSAR. As you continue to learn and apply these concepts, remember that persistence and continuous learning are key to mastering this powerful standard. Stay curious and keep exploring, and you'll be well on your way to becoming an AUTOSAR expert.

Get This Book's PDF Version and Exclusive Extras

UNLOCK NOW

Scan the QR code (or go to `packtpub.com/unlock`). Search for this book by name, confirm the edition, and then follow the steps on the page.

Note: Keep your invoice handy. Purchases made directly from Packt don't require one.

Index

A

action lists 193
actuator SWC 61
adaptive cruise control (ACC) 17
Adaptive Platform (AP) 15
administrative block 187
adt_VehSpeed 68
advanced driver assistance
    systems (ADAS) 4, 26, 46, 113, 207
analog-to-digital converters (ADCs) 34, 84
anti-lock braking system (ABS)
    control module 5
application data types 67
application layer 25, 26
application level 67
application software (ASW) 12
application/SWC software development 48
ARXML modeling 96, 97
assembly connector 74
    versus delegation connector 75
authoring tools 48, 69
automotive industry
    automotive software development 6
    ECU 5
    evolution 3-5
    traditional software development 7

automotive security 163
    advantages 164
    fundamentals 164, 165
    safety 166
    security, in AUTOSAR 165
automotive software development 6
automotive software standardization
    evolution 8
Autonomous Parking (AP) 214
autonomous parking assist system
    Application layer (SWCs) 217, 218
    AUTOSAR BSW modules 218
    development 214-217
AUTOSAR 23
    updates 213, 214
    URL 211
AUTOSAR authoring tool 69-72
AUTOSAR BSW modules 218
    communication stack 218
    crypto stack 218
AUTOSAR data types
    application data types 67
    base data types 68
    exploring 67, 69
    implementation data types 68

AUTOSAR documents
  accessing 210
  Explanation (EXP) 212
  principles 213
  Requirement Specification (RS) 212
  Software Specification (SWS) 211
  Template Protocol Specification (TPS) 212
AUTOSAR framework 9
  ECU, developing 13
  goals, achieving 12, 13
  impact, on traditional software
    development 11, 12
  technical goals 9
AUTOSAR methodology 43, 44
  code generation 45-50
  conformance classes 54
  ECU configuration 45-49
  system configuration 45-47
AUTOSAR OS
  architecture 115
  configuring 114, 115
  OSEK 113
  overview 111, 112
  real-time operating systems 112, 113
AUTOSAR OS architecture 115
  alarms 115
  events 116, 120, 122
  hooks 129
  interrupts 119, 120
  OS resources 128, 129
  resources 116
  scheduling 116, 122-124
  services 116
  tasks 115, 116
  task trigger mechanisms 124
AUTOSAR standard 13, 14
  Acceptance tests (ATs) 16
  AP 15

  Application interfaces (AIs) 17
  associate partners 11
  AUTOSAR platform 14
  core partners 10
  development partners 11
  Foundation (FO) 15
  premium partners 11
AUTOSAR SWCs
  application SWC 62
  complex device driver component 62
  NVBlock SWC 63
  parameter SWC 62
  service SWC 62
AUTOSAR templates 51
AUTOSAR UML Profile
  Basic Software Module
    Description Template 52
  ECU Resource Template 52
  Software Component Template 52
  System Template 52
AUTOSAR XML (ARXML) 52, 62

B

base data types 68
basic software (BSW) 63, 112, 200
  ECU abstraction layer 19
  Microcontroller abstraction
    layer (MCAL) 19
  services layer 19
**Basic Software Module
  (BswM) 136, 183, 192**
  use case, vehicle shutdown process 196, 197
basic tasks 116, 117
  ready 117
  running 117
  suspended 117
battery management system (BMS) 5, 25

block types
  managing 189
block types, storage objects
  dataset NVRAM block type 188, 189
  native NVRAM block type 188
  redundant NVRAM block type 188
BMS ECU
  application layer 40
  BSW, ECU abstraction layer 41
  BSW, MCAL 41
  BSW, service layer 40
  developing 39
  requirements 39, 40
  RTE 40
body control module (BCM) 5
BSW services
  communication services 32
  diagnostic services 32
  memory services 32
  system services 33
BUS State Manager (BUS-SM) 136

C

CAN Driver APIs 154
CAN Driver module 153
CanIf APIs 152
CanIf module
  features 151
CAN Interface (CanIf) 137
CAN modules 151
  CAN Driver APIs 154
  CAN Driver module 153
  CanIf APIs 152
  CanIf module 151
  CAN message sending example 154-157
CAN Network Management (CanNm) 150

CAN network states
  Bus-Off 158
  Bus-Sleep 158
  Normal Operation 158
  Prepare Bus-Sleep 158
  Ready Sleep 158
CanNm module 158
CAN State Manager (CANSM) 136, 196
**CAN Time Synchronization
    (CanTSyn) 206, 209, 210**
CAN Transceiver Driver (CanTrcv) 150
**CAN Transport
    Protocol (CanTp) 138, 151, 202**
Classic Platform (CP) 14
clean architecture 7
client-server communication 95
  ARXML modeling 96, 97
  asynchronous 99, 100
  communication types 97-100
  synchronous 98
client-server interface 65, 87
  AC system control 66
  client request 66
  climate analysis 66
  server operation 66
  user input 66
code generation 49, 50
COM module 141
  APIs 148, 149
  features 141, 142
  processing modes 147
  reception, handling 142
  signal filtering 146, 147
  transfer properties, for signals
      and signal groups 145
  transmission modes 143, 144
  transmissions, handling 142

communication channels, ComM
  full communication mode 157
  no communication mode 157
  silent communication mode 157
Communication (COM) module 177
Communication Manager (ComM) 136, 196, 211
communication matrix 47
communication model 83
  client-server communication 95
  information flow 84-87
  sender-receiver communication 87
  temperature sensor design 100
communication stack 133
communication types 91
  explicit SR communication 93, 94
  implicit SR communication 91, 92
  use case for Explicit read 94
  use case, for Implicit Read 93
complex device drivers (CDDs) 36
complex drivers 36, 37
compositions 73
  benefits 73
COM processing modes
  deferred processing 147
  immediate processing 147
COM stack 134
  application layer 134, 136
  Communication Driver Layer 137
  Communication Service Layer 136
  communication stack 134
  data communication example from PduR 138-140
  ECU Abstraction Layer 136
  overview 134-137
  PCI 138
  PDU 138
  RTE layer 134, 136
  Service Data Unit (SDU) 138
  signal groups 138
  signals 137
configuration classes 50
  link time 50
  post-build time 50
  pre-compile time 50
conformance classes 54
connector types 74
  assembly connector 74
  delegation connector 74
Control Algorithm (ControlAlgo) 218
controller area network (CAN) 13, 26, 134, 174, 201
cooperative scheduling 123
counters 124
Cruise Control SWC (receiver) 65
Crypto Interface (CryptoIf) 169
crypto module 171, 172
  hardware crypto 172, 173
  software crypto 172
crypto service manager module 169
  functions 169
  job concepts, in CSM 169, 170
  process flow 170, 171
C/S interfaces 95

D

data elements 88
data exchange
  templates, using for 51-53
DataReceiveEvent events 109
Default Error Tracer (DET) 204
deferred processing 147
delegation connector 74
  versus assembly connector 75

Diagnostic Communication
    Manager (DCM) 32, 174, 192
  features 201
Diagnostic Event
    Manager (DEM) 32, 202, 204
  responsibilities 202, 203
diagnostics
  Default Error Tracer (DET) 204
  Diagnostic Communication
    Manager (DCM) 202
  Diagnostic Event Manager 202
  Failure Management (FiM) 205
  overview 200
diagnostic system
  usage areas 200
diagnostic trouble codes (DTCs) 184, 201
digital input/output (DIO) 34
Direct Transmission Mode
  using, for triggered transfer 145

E

ecosystem components
  OEMs 10
  semiconductor manufacturers 10
  standard software vendors 10
  Tier 1 suppliers 10
ECU abstraction layer 33
  features 34
ECU configuration 45, 48, 49
ECU extract 48
ECU State Manager (ECUM) 33, 196
EEPROM Abstraction (EA) 185
electrically erasable programmable
    read-only memory (EEPROM) 184
electronic brake system (EBS) 188

electronic control unit (ECU) 5, 6,
    43, 44, 60, 111, 133, 176, 184
  developing, in AUTOSAR framework 13
  examples 5
electronic stability control (ESC) 4
engine control module (ECM) 5, 52
event handlers 103
events 102, 103, 120
  for temperature monitoring
    example 108, 109
  interacting, with interfaces
    and events 103, 104
  manipulating, with API functions 121, 122
  owning task 120
  unique identifier 120
extended tasks 118
  blocked state 118
  waiting state 118

F

Failure Management (FiM) 205
Firmware Over-the-Air (FOTA) 212
first in, first out (FIFO) stack 124
Flash Driver (FLS) 35
Flash EEPROM Emulation (FEE) 185
FlexRay Interface (FrIf) 137

G

gatewaying 140
General Purpose Input/Output (GPIOs) 52
Generic Structure 52

H

hashes 168
highest priority first mechanism 124

hooks 129
　error hooks 130
　shutdown hooks 129
　task/ISR execution hooks 129
human-machine interface (HMI) 18

I

immediate processing 147
implementation data types 68
　float32 68
　uint32 68
implementation level 67
Input/Output Hardware Abstraction 34
integration phase 114
Inter-ECU communication 72
interfaces 37, 64
　AUTOSAR interfaces 38
　interacting, with ports and events 103, 104
　internal behavior 67
　runnable entities 67
　sender-receiver 64
　standardized AUTOSAR interfaces 38
　standardized interfaces 38
interrupts 119, 120
　ISR category 1 and 2 119
interrupt service routine (ISR) 103, 116, 154
Intra-ECU communication 73
Intrusion Detection Reporter (IdsR) 174
Intrusion Detection System (IDS) 174
Intrusion Detection System Manager (IdsM) 174

K

Key Management (KeyM) 173

L

layer 17
　application layer 18
　AUTOSAR RTE 18
　basic software 19
layered software architecture 24, 25
　application layer (upper floor) 24
　basic software (BSW) layer (basement) 24
　runtime environment (ground floor) 24
LIN Interface (LinIf) 137
local interconnect network (LIN) 134

M

mask 120
MCU
　replacing 9
Memory Abstraction Interface (MemIf) 186
memory stack 184, 185
　abstraction layer 185
　MCAL 186
　non-volatile memory, significance 184
　service layer 185
　storage objects 186
　use case 190, 192
Message Authentication Code (MAC) 169
microcontroller abstraction layer (MCAL) 34-36, 186
microcontroller drivers
　exchanging 9
mode management 33, 192
　BSWM use case 196, 197
　role and functionality 193-195

N

name 120

Network Management (Nm) 135, 157
  key modules 157-159
Non-Volatile Memory (NVM) 171, 183
non-volatile random access memory (NVRAM) Manager (NvmM) 185
NV block 187
NvM Block Descriptor
  configuration parameters 189
NVRAM blocks 186

O

Object Detection (ObjDetect) 217
Onboard Diagnostics (OBD) 203
operating system (OS) 24, 77
original equipment manufacturer (OEM) 7, 38, 45
OS resources 128, 129

P

PCI 138
PDU handling 141
Periodic Transmission Mode
  used, for pending transfer 146
ports 63
  interacting, with interfaces and events 103, 104
  Provided Ports (PPorts) 63
  Required Ports (RPorts) 64
post-build time parameters
  loadable 50
  selectable 50
printed circuit board (PCB) 5
priority inversion 113
processing modes, CAN Driver
  interrupt mode 153
  polling mode 153

Protocol Control Information (PCI) 138
Protocol Data Unit (PDU) 136
Protocol Data Unit Router (PDUR) 137, 178, 202
  features 149, 150
Provided Ports (PPorts) 63
pulse-width modulation (PWM) 34

R

radar ECUs
  synchronizing, with StbM 208
radio frequency (RF) 208
RAM block 187
read-only memory (ROM) 180
real-time operating system (RTOS) 14, 112, 113
relative schedule tables 126
Remote Procedure Calls (RPCs) 28
Repeat Message State 159
request-response model 65
Required Ports (RPorts) 64
ROM block 187
Rte_Call_TempInputPort_GetTemperature function 71
RTE events
  DataReceivedEvent 107
  OperationInvokedEvent 106
  TimingEvent 105
  types 104-108
runtime environment (RTE) 12, 27, 83, 193
  features 76
  function 28-30
  RTE functions 28-30
  significance 76, 77
  VFB concept 27, 28

runtime environment (RTE) generation 77
  output 79-81
  phases 78
  tools 78
  tools, selecting 79
runtime environment (RTE) generation, input example
  ECU configuration 77
  SWC descriptions 77
  Toolchain 77

S

safety 166
schedule tables 126
scheduling 122
scheduling policies
  full preemptive scheduling 122
  non-preemptive scheduling 123
secure boot 180
Secure Onboard Communication (SecOC) 169
security 165, 166
Security Event Memory (SEM) 174
security features 167
security stack architecture 167
  cryptographic techniques 167, 169
  crypto interface module 171
  crypto module 171
  crypto service manager module 169
  features 167
  helper modules 173
sender-receiver communication
  AUTOSAR XML (ARXML) modeling 89-91
  communication types 91
  Receiver (R Port) 87
  sender (P Port) 87

sender-receiver interface 64, 87
  Cruise Control SWC (receiver) 65
  Speed Display SWC (receiver) 64
  Speed Sensor SWC (sender) 64
sensor SWC 61
sequential access 124
Service Data Unit (SDU) 138
service-oriented architecture (SOA) 15
signal 134, 137
signal grouping 141
signal groups 138, 141
signal handling 141
software architecture and design 17
  layer 18
  stack 19
software components (SWCs) 7, 23, 59, 60, 200
  AUTOSAR SWCs types 62
  characteristics 60
  communication between 72, 73
  compositions 73
  elements 63, 66, 67
  modeling 69
  throttle control component 61, 62
software components (SWCs), elements
  interfaces 64
  ports 63
software layer 31
  complex drivers 36, 37
  ECU abstraction layer 33, 34
  microcontroller abstraction layer 34-36
  service layer 31, 32
Software Specification (SWS) 211
Speed Display SWC (receiver) 64
Speed Sensor SWC (sender) 64
stack 19, 20
stack monitoring tool 113
standard architecture 10

Steering Control (SteeringCtrl) 218
storage objects, memory stack 186, 187
  block properties 189, 190
  block types 188
SWC Description file 48
SW modules
  ADCDrv 86
  ADC reading module 85
  CANDrv 86
  Display SW module 86
  temp conversion module 85
synchronized schedule tables 126
Synchronized Time-Base Manager (StbM) module 206, 207
  radar ECUs, synchronizing with StbM 208
system configuration 45, 46, 47
System Configuration Description 47
system under test (SUT) 16

T

tasks
  basic tasks 116, 117
  extended tasks 118
  hooks 116
task state model 117
task trigger mechanisms 124
  alarms 124-126
  schedule tables 126, 127
TemperatureHandler component 70
temperature sensor design 100-102
Template Protocol Specification (TPS) 212
throttle control component 61
  functions 61
Throttle Control (ThrottleCtrl) 218
tick 124
Time Master 206

Time Slave 206
time synchronization
  CAN Time Synchronization (CanTSyn) 209, 210
  overview 206, 207
  significance 207
timing protection 124
traditional software development 7, 8
  limitations 7
transmission control module (TCM) 5
transmission modes, I-PDUs
  DIRECT (N-Times) 143
  mixed 144
  none 144
  periodic 143

U

unified diagnostic services (UDS) 39, 201
Unified Modeling Language (UML) 51
use cases and examples, Crypto Stack
  data encryption 174-176
  secure boot 180, 181
  secure communication 177, 178
  secure diagnostics 179
User Interface Display (UIDisplay) 218

V

Vehicle Speed (VehSpd) 136, 137
virtual functional bus (VFB) 27, 28, 72

W

waiting state 118

www.packtpub.com

Subscribe to our online digital library for full access to over 7,000 books and videos, as well as industry leading tools to help you plan your personal development and advance your career. For more information, please visit our website.

Why subscribe?

- Spend less time learning and more time coding with practical eBooks and Videos from over 4,000 industry professionals
- Improve your learning with Skill Plans built especially for you
- Get a free eBook or video every month
- Fully searchable for easy access to vital information
- Copy and paste, print, and bookmark content

Did you know that Packt offers eBook versions of every book published, with PDF and ePub files available? You can upgrade to the eBook version at packtpub.com and as a print book customer, you are entitled to a discount on the eBook copy. Get in touch with us at customercare@packtpub.com for more details.

At www.packtpub.com, you can also read a collection of free technical articles, sign up for a range of free newsletters, and receive exclusive discounts and offers on Packt books and eBooks.

Other Books You May Enjoy

If you enjoyed this book, you may be interested in these other books by Packt:

Industrial Automation from Scratch

Olushola Akande

ISBN: 978-1-80056-938-6

- Get to grips with the essentials of industrial automation and control
- Find out how to use industry-based sensors and actuators
- Know about the AC, DC, servo, and stepper motors
- Get a solid understanding of VFDs, PLCs, HMIs, and SCADA and their applications
- Explore hands-on process control systems including analog signal processing with PLCs
- Get familiarized with industrial network and communication protocols, wired and wireless networks, and 5G
- Explore current trends in manufacturing such as smart factory, IoT, AI, and robotics

Internet of Things from Scratch

Renaldi Gondosubroto

ISBN: 978-1-83763-854-3

- Gain a holistic understanding of IoT basics through real-life use cases
- Explore communication protocols and technologies integral to IoT
- Use AWS to build resilient, low-latency networks
- Construct complex IoT networks, building upon foundational principles
- Integrate data analytics workloads and generative AI seamlessly with IoT
- Understand the security threat landscape of IoT and how to mitigate these risks
- Develop industry-grade projects within the open source IoT community
- Embrace a futuristic perspective of IoT by understanding both risks and rewards

Packt is searching for authors like you

If you're interested in becoming an author for Packt, please visit `authors.packtpub.com` and apply today. We have worked with thousands of developers and tech professionals, just like you, to help them share their insight with the global tech community. You can make a general application, apply for a specific hot topic that we are recruiting an author for, or submit your own idea.

Share Your Thoughts

Now you've finished *AUTOSAR Fundamentals and Applications*, we'd love to hear your thoughts! Scan the QR code below to go straight to the Amazon review page for this book and share your feedback or leave a review on the site that you purchased it from.

`https://packt.link/r/1-805-12087-5`

Your review is important to us and the tech community and will help us make sure we're delivering excellent quality content.

www.ingramcontent.com/pod-product-compliance
Lightning Source LLC
Chambersburg PA
CBHW080836230426
43665CB00021B/2860